Intuition, Trust, and Analytics

Data Analytics Applications

Series Editor: Jay Liebowitz

Intuition, Trust, and Analytics

Edited by
Jay Liebowitz
Joanna Paliszkiewicz
Jerzy Gołuchowski

CRC Press
Taylor & Francis Group
Boca Raton London New York

CRC Press is an imprint of the
Taylor & Francis Group, an **informa** business

CRC Press
Taylor & Francis Group
6000 Broken Sound Parkway NW, Suite 300
Boca Raton, FL 33487-2742

First issued in paperback 2022

ISBN 13: 978-1-03-247651-3 (pbk)
ISBN 13: 978-1-1387-1912-5 (hbk)

DOI: 10.1201/9781315195551

This book contains information obtained from authentic and highly regarded sources. Reasonable efforts have been made to publish reliable data and information, but the author and publisher cannot assume responsibility for the validity of all materials or the consequences of their use. The authors and publishers have attempted to trace the copyright holders of all material reproduced in this publication and apologize to copyright holders if permission to publish in this form has not been obtained. If any copyright material has not been acknowledged please write and let us know so we may rectify in any future reprint.

Publisher's Note
The publisher has gone to great lengths to ensure the quality of this reprint but points out that some imperfections in the original copies may be apparent.

Library of Congress Cataloging-in-Publication Data

Names: Liebowitz, Jay, 1957- editor. | Paliszkiewicz, Joanna Olga, editor. | Gołuchowski, Jerzy, editor.
Title: Intuition, trust, and analytics / edited by Jay Liebowitz, Harrisburg University of Science and Technology, Joanna Paliszkiewicz, Warsaw University of Life Sciences, SGGW, Jerzy Goluchowski, University of Economics in Katowice.
Description: Boca Raton, Florida : CRC Press, [2018]
Identifiers: LCCN 2017018340| ISBN 9781138719125 (hardback) | ISBN 9781315195551 (e-book) | ISBN 9781351764407 (e-book) | ISBN 9781351764391 (e-book) | ISBN 9781351764384 (e-book)
Subjects: LCSH: Decision making. | Intuition. | Trust.
Classification: LCC HD30.23 .I59 2018 | DDC 658.4/03—dc23
LC record available at https://lccn.loc.gov/2017018340

Visit the Taylor & Francis Web site at
http://www.taylorandfrancis.com

and the CRC Press Web site at
http://www.crcpress.com

My intuition tells me that I better thank my family first, as well as my students, colleagues, and friends along the way. It has also been a pleasure working on this book with such wonderful friends—Joanna, Jerzy, and John (the publisher). (*JL*)

I always trust my heart and intuition, no one can give you wiser advice than yourself.

This book is dedicated to:

- My family: My mother Anna, and in memory of my father Jerzy, my husband Radek, and my daughter Natalia—who always support me
- The Mentors: Jay and Jerzy—who always inspire me
- And John, the publisher—who helped us to prepare this book (*JP*)

I'd like to give special thanks to my family, especially my dearest mother and my beloved wife Regina, for years of continuous support, my colleagues, and last but not least Joanna, Jay, and John without whom this book would not see the light of day. (*JG*)

Contents

SECTION III   ANALYTICS

Preface

How Intuition, Trust, and Analytics Play a Role in Executive Decision-Making

Jay Liebowitz
Harrisburg University of Science and Technology
Joanna Paliszkiewicz
Warsaw University of Life Sciences—SGGW
Jerzy Gołuchowski
University of Economics in Katowice

Executive Decision-Making

Executives and managers must make decisions in their work. Some of these decisions are made spontaneously, based on intuition, and others may need to apply data-driven approaches through analytics and Big Data. Intuition has a significant role in trust-building in business (Parikh, Neubauer, & Lank, 1994; Burke & Miller, 1999; Hayes, Allinson, & Armstrong, 2004; Sadler-Smith & Shefy, 2004; Miller & Ireland, 2005; Dane & Pratt, 2007; Salas, Rosen, & Diaz Granados, 2010; Miles & Sadler-Smith, 2014; Pretz & Liebowitz, 2016). By using intuition, for example, people can develop trust in themselves, which creates the foundations to develop trust to others. Many researchers and practitioners have explored the significance of trust in: the organization (Cacioppo et al. 1984; Mayer, Davis, & Schoorman, 1995; Lewicki & Bunker, 1995; Bibb & Kourdi, 2004; Six, 2004; Sprenger, 2004; Paliszkiewicz, 2013), marketing (Soh, Reid, & King, 2007; Lee & Rao, 2010; Cho, Huh, & Faber, 2014), social media (Paliszkiewicz & Koohang 2016), Corporate Social Responsibility (CSR) (Garbarino & Johnson, 1999; Gołuchowski & Losa-Jonczyk, 2013), and in knowledge management processes especially in knowledge sharing and organizational performance (Koohang, Paliszkiewicz, & Gołuchowski, 2017).

In this age of "Big Data," knowledge gained from experiential learning may take a back seat to analytics. However, the use of intuition and trust in executive decision-making should play an important role in the decision process. In fact, a KPMG

(2016a) study found that just one-third of CEOs trust data analytics, mainly due to concerns about internal data quality. KPMG (2016b) in their second study also found that most business leaders believe in the value of using data and analytics, but say they lack confidence in their ability to measure the effectiveness and impact of data and analytics, and mistrust the analytics used to help drive decision-making. Unfortunately, in the data analytics community, intuition typically hasn't been discussed in terms of its application in executive decision-making.

Trusting Your Intuition

In looking at some examples of the research in this area, Kandasamy et al. (2016) found that hedge fund traders who relied on their gut feelings outperformed those who didn't. Wang, Highhouse, Lake, Petersen, and Rada (2017) found that intuition and analysis are independent constructs, rather than opposite ends of a bipolar continuum. Liebowitz (2014) discusses both the research and application of applying intuition-based decision-making across various sectors. Hanlon (2011) discussed how managers rationalize intuition in their strategic decision-making. Usher, Russo, Weyers, Brauner, and Zakay (2011) demonstrate how intuition can be used in place of analytics for complex decision-making. Swami (2013) found that in situations involving higher time pressure, higher stakes, or increased ambiguities, experts may well use intuitive decision-making rather than structured approaches. Violino (2014) indicates a study by the Economist Intelligence Unit and Applied Predictive Technologies that found nearly three-quarters of the executives surveyed say they trust their own intuition when it comes to decision-making, and 68 percent believe they would be trusted to make a decision that was not supported by data. Akinci and Sadler-Smith (2012) provide a thorough historical review of intuition in management research. Woiceshyn (2009) has shown that effective CEOs share three thinking-related traits: focus, motivation, and self-awareness. Moxley, Ericsson, Charness, and Krampe (2012) found that both experts and less skilled individuals benefit significantly from extra deliberation regardless of whether the problem is easy or difficult. According to Hassani, Abdi, and Jalali (2016), intuition is a way of learning and is a type of legitimate knowledge in nursing. Certainly, intuition-based decision-making is an increasingly popular topic among those in the Naturalistic Decision-Making, Fast and Frugal Heuristics, and Heuristics and Biases communities (Klein, 2015).

Intuition Surveys of Caribbean and U.S. C-Level Executives

To gain further insight into how executives may be using intuition, Liebowitz applied Pretz's et al. (2014) Intuition Scale survey with two groups of C-level executives in 2016 during separate workshops. One group was Caribbean executives, and the other group comprised U.S. executives. The Caribbean sample had 29 respondents; the U.S. sample had 27 respondents. Tables 1 and 2 show the results of the surveys.

Holistic intuitions are judgments based on a qualitatively non-analytical process. As the names imply, Holistic–Big Picture looks at the "big picture" view, and Holistic–Abstract looks more at the abstract view. Affective intuitions are judgments based mainly on emotional reactions to decision situations. Inferential intuitions are judgments based on automated inferences, decision-making processes that were once analytical but have become intuitive in practice (Pretz et al., 2014; Pretz & Liebowitz, 2016).

Table 1 shows the mean scores for each question for the Caribbean executives and for the U.S. executives. Generally speaking, the U.S. scores were higher, meaning that their responses tended to fall more toward the neutral to true values versus those mean averages for the Caribbean executives. However, there were some interesting exceptions. For example, the question about "Is it better to break a problem into parts than to focus on the big picture?" showed that the mean value for the Caribbean executives was higher than that for the U.S. executives, indicating that the Caribbean executives seemed to prefer logic to emotions. The U.S. executives indicated that they seemed to trust their intuition in their area of expertise more than the Caribbean executives.

Table 2 provides the mean values for the types of intuition based on Pretz et al. (2014). Based on Table 2, it appears that the U.S. executives have higher scores for each of the four types of intuition, as compared with those scores from the Caribbean executives. The U.S. executives have scores between neutral and mostly true, whereas the Caribbean executives have scores between mostly false and neutral. It appears that the U.S. executives generally have a more positive outlook in terms of their use of intuition in their decision-making. Of course, the sample sizes for this experiment were small (29 for the Caribbean executives; 27 for the U.S. executives), so it is difficult to generalize. However, this experiment might suggest future research looking at cultural issues in terms of how intuition plays a role in executive decision-making.

Table 1 Mean Scores on the Intuition Scale Survey of Caribbean and U.S. C-Level Executives

Question Number	Caribbean Mean Response	U.S. Mean Response
1	2.79	4.07
2	4.07	4.41
3	2.10	2.56
4	3.76	4.15
5	3.62	3.41

(*Continued*)

Table 1 (*Continued*)   Mean Scores on the Intuition Scale Survey of Caribbean and U.S. C-Level Executives

Question Number	Caribbean Mean Response	U.S. Mean Response
6	4.00	4.00
7	3.72	3.48
8	3.59	4.18
9	2.41	2.70
10	3.52	4.00
11	2.24	2.41
12	3.89	4.15
13	2.35	2.70
14	2.03	2.07
15	3.14	3.33
16	3.35	3.89
17	3.62	4.15
18	3.76	3.93
19	4.31	4.22
20	2.79	3.19
21	3.24	3.74
22	3.93	4.15
23	3.38	3.52
24	2.45	2.93
25	2.21	1.93
26	3.93	3.70
27	3.97	4.19
28	3.10	2.82
29	3.52	3.82

Source: Pretz, J. et al., *J. Behav. Decis. Mak.*, 27, 2014.

Note: 1 = Definitely False; 2 = Mostly False; 3 = Undecided (neither true nor false); 4 = Mostly True; 5 = Definitely True.

Table 2 Mean Scores of Different Types of Intuition Using the Caribbean C-Level Executives versus the U.S. C-Level Executives

Type of Intuition	Caribbean Executives	U.S. Executives
H-BP	3.18	3.41
I	3.35	3.94
HA	2.82	3.09
A	2.93	3.13

Source: Pretz, J. et al., *J. Behav. Decis. Mak.*, 27, 2014.

Note: H-BP = Holistic Intuition–Big Picture; I = Inferential Intuition; HA = Holistic Intuition–Abstract; A = Affective Intuition; 1 = Definitely False; 2 = Mostly False; 3 = Undecided; 4 = Mostly True; 5 = Definitely True.

Using Intuition for Decision-Making

To get a sense for how those interested in data analytics view intuition in their decision-making, a web-based survey was sent to the attendees of the Third Data Analytics Summit held on March 9–10, 2017 at Harrisburg University of Science and Technology. Screenshots from the first 100 respondents are shown below. The results show that intuition is important, but certainly "rational intuition" may be the way to go for decision-making to back up one's hunches with data and analysis. Of course, the sample was biased as most of the Data Analytics Summit attendees were data-driven oriented (see next page).

Improving Intuition Awareness for Executives

Intuition is something that can be honed. Through experiential learning and techniques to further stimulate one's intuitive sense, intuition can play an important role in executive decision-making. According to Davis-Floyd and Arvidson (1997), exercises have been used for many years to further develop one's intuition. Even organizations have held training courses to instill a greater intuitive capacity in their employees. For example, the Marine Corps stresses intuition as a central key to victory. Bell Atlantic Corporation listed intuition as an important quality on job descriptions. Even among midwives, intuition is a salient course of authoritative knowledge (David-Floyd & Arvidson, 1997). Sadler-Smith and Shefy (2004) highlight some practical guidelines to becoming an "intuitive executive." They include (Sadler-Smith & Shefy, 2004):

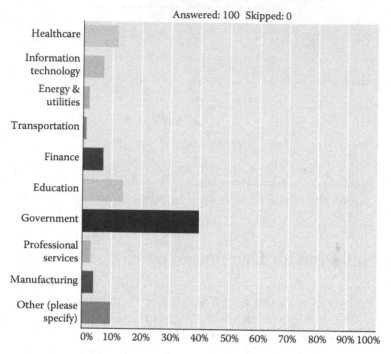

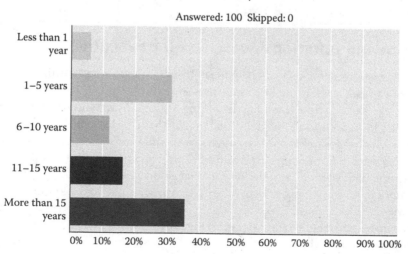

I trust my intuitions, especially in familiar
situations.

Answered: 100 Skipped: 0

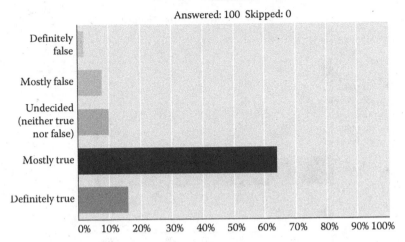

I prefer to use my emotional hunches to
deal with a problem, rather than thinking
about it.

Answered: 100 Skipped: 0

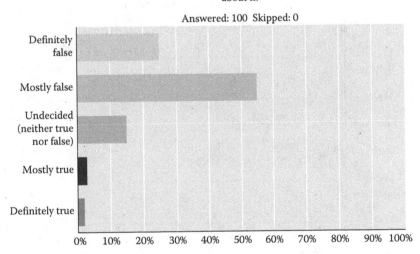

There is a logical justification for most of
my intuitive judgments.

Answered: 100 Skipped: 0

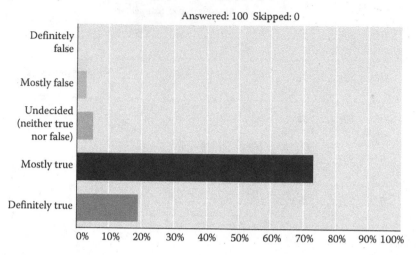

When tackling a new project, I concentrate
on big ideas rather than the details?

Answered: 100 Skipped: 0

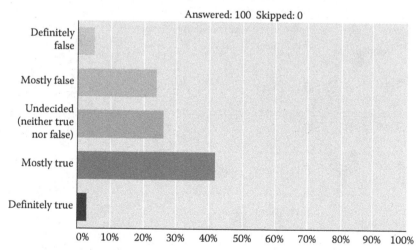

I tend to use my heart as a guide for my actions.

Answered: 100 Skipped: 0

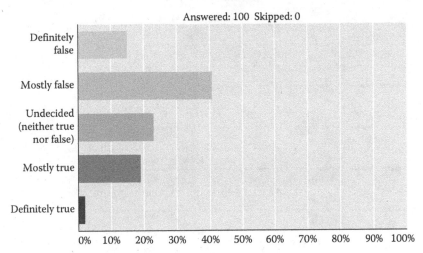

I would rather think in terms of theories than facts.

Answered: 100 Skipped: 0

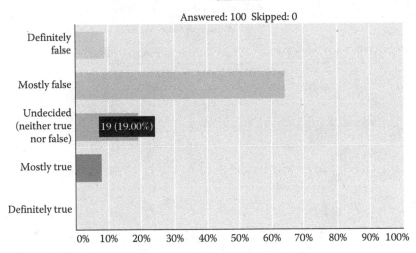

My approach to problem solving relies
heavily on my past experience.

Answered: 100 Skipped: 0

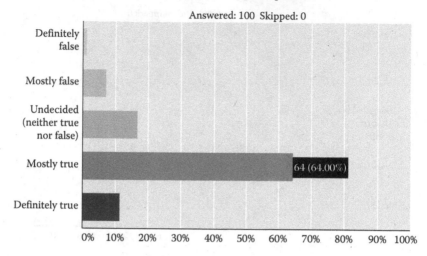

When making decisions, I value my feelings
and hunches just as much as I value facts.

Answered: 100 Skipped: 0

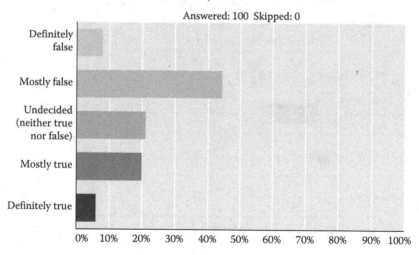

Open the closet: To what extent do you: experience intuition; trust your feelings; count on intuitive judgments; suppress hunches; covertly rely upon gut feel?

Don't mix up your I's: Instinct, insight, and intuition are not synonymous; practice distinguishing between your instincts, your insights, and your intuitions.

Elicit good feedback: Seek feedback on your intuitive judgments; build confidence in your gut feel; create a learning environment in which you can develop better intuitive awareness.

Get a feel for your batting average: Benchmark your intuitions; get a sense for how reliable your hunches are; ask yourself how your intuitive judgment might be improved.

Use imagery: Use imagery rather than words; literally visualize potential future scenarios that take your gut feelings into account.

Play devil's advocate: Test out intuitive judgments; raise objections to them; generate counterarguments; probe how robust gut feel is when challenged.

Capture and validate your intuitions: Create the inner state to give your intuitive mind the freedom to roam; capture your creative intuitions; log them before they are censored by rational analysis.

These suggestions may increase one's intuitive awareness, as well as even doing meditation and introspection. There is also a "spirituality management" movement which suggests that decisions may be influenced from "on-high" (in this case, way above the CEO and Chairman of the Board).

Of course, even though this preface looks at intuition, analytics and trust should still be part of the winning formula for making sound executive decisions. The adage that "Intuition + Analytics + Trust = Success" should be a great combination for any kind of decision-making.

This preface advocates that intuition should not be disregarded when making executive decisions. The knowledge management discipline shows us that experiential learning adds greatly to the innovation process. Analytics, as a data-driven approach, can also add value and insights to the decision-making process. Unfortunately, the "Analytics" community and the "Judgment and Decision-Making" community haven't quite synergized. The hope is that this book, in general, will provide reasons why these two communities should be united, along with those in the "Trust" community, to improve strategic decision-making in organizations and beyond.

References

Akinci, C., & Sadler-Smith, E. (2012). Intuition in management research: A historical review. *International Journal of Management Reviews, 14*, British Academy of Management.

Bibb, S., & Kourdi, J. (2004). *Trust matters for organizational and personal success.* New York: Palgrave Macmillan.

Burke, L. A., & Miller, M. K. (1999). Taking the mystery out of intuitive decision making. *Academy of Management Executive*, 13(4), 91–99.

Cacioppo, J. T., Petty, R. E., & Kao, C. F. (1984). The efficient assessment of need for cognition. *Journal of Personality Assessment, 48*, Taylor & Francis.

Cho, S., Huh, J., & Faber, R. J. (2014). The influence of sender trust and advertiser trust on multistage effects of viral advertising. *Journal of Advertising*, 43(1), 100–114.

Dane, E., & Pratt, M. (2007). Exploring intuition and its role in managerial decision making. *Academy of Management Review*, 32(1), 33–54.

Davis-Floyd, R., & Arvidson, P. S. (Eds.) (1997). *Intuition: The inside story, interdisciplinary perspectives.* Princeton University, Princeton Engineering Anomalies Research Laboratory. New York and London: Routledge.

Garbarino, E., & Johnson, M. S. (1999). The different roles of satisfaction, trust, and commitment in customer relationships. *The Journal of Marketing*, 63(2), 70–87.

Gołuchowski, J., & Losa-Jonczyk, A. (2013). Wykorzystanie nowych mediów w promocji idei społecznej odpowiedzialności uczelni. [Using new media for the promotion of university social responsibility]. *Studia Ekonomiczne*, 13(157), 67–78.

Hanlon, P. (2011). The role of intuition in strategic decision making: How managers rationalize intuition. *14th Annual Conference of the Irish Academy of Management Proceedings*, Dublin, August 31-September 2.

Hassani, P., Abdi, A., & Jalali, R. (2016). State of Science, "Intuition in Nursing Practice": A Systematic Review Study. *Journal of Clinical & Diagnostic Research*, 10(2), JE07–JE11.

Hayes, J., Allinson, C. W., & Armstrong, S. J. (2004). Intuition, women managers and gendered stereotypes. *Personnel Review*, 33(4), 403–417.

Kandasamy, N., Garfinkel, S., Page, L., Hardy, B., Critchley, H., Gurnell, M., & Coates, J. (2016). Interoceptive ability predicts survival on a London trading floor. Scientific Reports, University of Cambridge, September 19.

Klein, G. (2015). A naturalistic decision making perspective on studying intuitive decision making. *Journal of Applied Research in Memory and Cognition, 4*(3), 164–168.

Koohang, A., Paliszkiewicz, J., & Gołuchowski J. (2017). The impact of leadership on trust, knowledge management, and organizational performance: A research model. *Industrial Management & Data Systems*, 117(3), 521–537.

KPMG. (2016a). KPMG study: Just one-third of CEOs trust aata analytics. July, http://solutionsreview.com/business-intelligence/kpmg-study-just-one-third-of-ceos-trust-data-analytics/.

KPMG. (2016b). Building trust in analytics. October 31.

Lee, S., & Rao, V. S. (2010). Color and store choice in electronic commerce: The explanatory role of trust. *Journal of Electronic Commerce Research*, 11(2), 110–126.

Lewicki, R. J., & Bunker, B. B. (1995). Trust in relationships: A model of trust development and decline. In B. B. Bunker & J. Z. Rubin (Eds.), *Conflict, cooperation and justice* (pp. 131–145). San Francisco, CA: Jossey-Bass.

Liebowitz, J. (Ed.). (2014). *Bursting the big data bubble: The case for intuition-based decision making.* Boca Raton, FL: Taylor & Francis.

Mayer, R. C., Davis, J. H., & Schoorman, F. D. (1995). An integrative model of organizational trust. *Academy of Management Review*, 20(3), 709–734.

Miles, A., & Sadler-Smith, E. (2014). With recruitment I always feel I need to listen to my gut: The role of intuition in employee selection. *Personnel Review*, 43(4), 606–627.

Miller, C. C., & Ireland, R. D. (2005). Intuition in strategic decision making: Friend or foe in the fast-paced 21st century? *Academy of Management Executive*, 19(1), 19–30.

Moxley, J., Ericsson, K., Charness, N., and Krampe, R. (2012). The role of intuition and deliberative thinking in experts' superior tactical decision making. *Cognition*, 124. Elsevier.

Paliszkiewicz, J. (2013). *Zaufanie w zarządzaniu*. [Trust in management]. Warszawa: Wydawnictwo Naukowe PWN.

Paliszkiewicz, J., & Koohang, A. (2016). Social media and trust: A multinational study of university students. Santa Rosa, CA: Informing Science Press.

Parikh, J., Neubauer, F., & Lank, A. G. (1994). *Intuition: The new frontier of management*. London: Blackwell.

Pretz, J., & Liebowitz, J. (2016). Executives in data analytics trust intuition over analytics. Research Poster Session, Society for Judgment and Decision Making, 37th Annual Conference, Boston.

Pretz, J., Brookings, J., Carlson, L., Humbert, T., Roy, M., Jones, M., & Memmert, D. (2014). Development and validation of a new measure of intuition: The types of intuition scale. *Journal of Behavioral Decision Making*, 27. John Wiley.

Sadler-Smith, E. & Shefy, E. (2004). The intuitive executive: understanding and applying 'gut feel' in decision making. *Academy of Management Executive*, 18(4), 76–91.

Salas, E., Rosen, M., & Diaz Granados, D. (2010). Expertise-based intuition and decision making in organizations. *Journal of Management*, 36(4), 941–973.

Six, F. (2004). Trust and trouble. Building interpersonal trust within organizations. Rotterdam: Erasmus Research Institute of Management.

Soh, H., Reid, L. N., & King, K. W. (2007). Trust in different advertising media. *Journalism and Mass Communication*, 84(3), 455–476.

Sprenger, R. K. (2004). Trust. *The best way to manage*. London: Cyan/Campus.

Swami, S. (2013). Executive functions and decision making: A managerial review. *IIMB Management Review*, 25(4), 203–212.

Usher, M., Russo, Z., Weyers, M., Brauner, R., & Zakay, D. (2011). The impact of the mode of thought in complex decisions: intuitive decisions are better. *Frontiers in Psychology*, 2, 1–13

Violino, B. (2014). Do executives trust data or intuition in decision making? *Information Management*. http://www.information-management.com.

Wang, Y., Highhouse, S., Lake, C., Petersen, N., & Rada, T. (2017). Meta-analytic investigations of the relation between intuition and analysis. *Journal of Behavioral Decision Making*, 30(1): 15–25.

Woiceshyn, J. (2009). Lessons from 'good minds': How CEOs use intuition, analysis and guiding principles to make strategic decisions. *Long Range Planning*, 42(3), 298–319.

Contributors

Stephen Adams
University of Virginia
Charlottesville, Virginia

Peter A. Beling
University of Virginia
Charlottesville, Virginia

Alan Briggs
SAS Institute
Ellicott City, Maryland

James Wing-Kin Cheng
Memorial Sloan Kettering Cancer
    Center
New York, New York

Viktor Dörfler
University of Strathclyde
Scotland, United Kingdom

Katarína Fichnová
Constantine the Philosopher University
    in Nitra
Nitra, Slovakia

Barbara Filipczyk
University of Economics in Katowice
Katowice, Poland

Jerzy Gołuchowski
University of Economics in Katowice
Katowice, Poland

Lorri A. Halverson
University of Sioux Falls
Sioux Falls, South Dakota

Judith Hurwitz
Hurwitz & Associates
Needham, Massachusetts

Dorota Konieczna
University of Economics in Katowice
Katowice, Poland

Alex Koohang
Middle Georgia State University
Macon, Georgia

Barbara Kożuch
Jagiellonian University
Cracow, Poland

Regina Lenart-Gansiniec
Jagiellonian University
Cracow, Poland

Jay Liebowitz
Harrisburg University of Science and
  Technology
Harrisburg, Pennsylvania

Anna Losa-Jonczyk
University of Economics in Katowice
Katowice, Poland

Peter Mikuláš
Constantine the Philosopher University
  in Nitra
Nitra, Slovakia

Lionel Page
Queensland University of Technology
Brisbane, Australia

Joanna Paliszkiewicz
Warsaw University of Life Sciences—
  SGGW
Warsaw, Poland

Wilds Ross
KPMG LLP
Washington, DC

William T. Scherer
University of Virginia
Charlottesville, Virginia

Natalia V. Smirnova
American Institute for Economic
  Research (AIER)
Great Barrington, Massachusetts

Marc Stierand
École hôtelière de Lausanne
HES-SO // University of Applied
  Sciences Western Switzerland
Lausanne, Switzerland

Łukasz P. Wojciechowski
University of Ss. Cyril and Methodius
Trnava, Slovakia

INTUITION

1

Chapter 1

The Underpinnings of Intuition

Viktor Dörfler and Marc Stierand

Contents

The intuitive mind is a sacred gift and the rational mind is a faithful servant. We have created a society that honors the servant and has forgotten the gift.

Albert Einstein

Understanding intuition puzzled many researchers. Only philosophers feel comfortable to think about intuition not only as legitimate, but also as a possibly superior form of knowledge (see, e.g., Bergson, 1911, 1946; Jung, 1921, § 770; Spinoza, 1677, Part 5). It was during this early stage of intuition research, that philosophy provided the basis for one of the most fundamental claims in the human studies; if we are to fully understand human consciousness, we must also understand intuition. In fact, as David Chalmers (1998, p. 110) argues, intuition is "the very raison d'être," why we know so little about human consciousness. Thus, psychologists started to develop the so-called "dual process theory" that later also found recognition within management and organization research. Although intuition has found its way into mainstream research, we cannot say that we have a widespread agreement about some fundamentals of intuition, that is, whether it can be ultimately reduced to firings of neurons, should it be regarded as a complex mental phenomenon, or whether we should regard it as something mystical. Of course, intuition is not fundamentally different from other mental phenomena, only due to its peculiar characteristics discussed below, the possibility of looking at it in different positions are more apparent. However, we believe that this lack of agreement will not prevent scholarly attempts to understand intuition better. Currently, researchers with very different beliefs seem to be able to build on one another's results, and work together in a joint endeavor to catch the essence of this particularly interesting and beautiful mental phenomenon.

Even though intuition as a valuable decision-making tool used by managers, particularly top executives, was deemed reasonable, it was not before Chester Barnard (1938) published his seminal book *The Functions of the Executive* that exploring intuition started. Although Chester Barnard was a practitioner himself, his book has been widely accepted in academia, and in the field of intuition it marked the beginning of scholarly interests in intuition. The first academic inquiry regarding intuition was Herbert Simon's work on *Administrative Behavior* (first edition published 1947). This led to one of the most cited descriptions of intuition: "Intuition and judgment—at least good judgment—are simply analyses frozen into habit" (Simon, 1987, p. 63). This description was followed by the study by Weston Agor (1986), the first empirical research on managers' intuition, in which he explored both successes and failures—this has a symbolic value, as it shows that those who argue for the importance of intuition are not blind to the failures of intuition.

In this chapter, we will portray intuition in the context of decision-making by combining understandings from a variety of areas and drawing on both practitioners as well as academic sources. We limit, for the sake of understanding, intuition to consist of intuitive knowledge that is often accompanied by somatic and affective charges, thus ignoring the multipotential aspect of intuition and generally, cognition (see, e.g., Dörfler & Szendrey, 2008). We will establish a link between intuition and different levels of expertise, but essentially, will focus on intuition at the highest level of expertise.

What Is Intuition?

Intuition isn't the enemy, but the ally, of reason.

John Kord Lagemann

Perhaps the easiest way to conceptualize intuiting is to see it as a way of "direct knowing," that is, knowing "without any use of conscious reasoning" (Sinclair & Ashkanasy, 2005, p. 357). Although this is not an all-encompassing explanation, most people seem to understand it intuitively, and thus we adopt it as our starting point. Direct knowing means that knowledge is not achieved by the step-by-step reasoning that typically characterizes the academic view of decision-making, but through a process that somehow seems to bypass these steps. Usually, when a concept is so vague, such as intuition, we often contrast it with something, that is, explain it through what it is not. Thus, intuition is often contrasted with analysis or with rational methods. However, none of these contrasts seem to stand scrutiny. The opposite of analysis is not intuition but synthesis. It is true that intuition often entails synthesis, and this is the point where Mintzberg (1994) challenged Simon on his conceptualization of intuition as being "analysis frozen into habit," arguing that intuition is about synthesis, which one will never achieve through analyses. Yet, synthesis can also be achieved through step-by-step reasoning, not only by means of intuiting. Similarly, intuition sometimes seems to bypass the analytical steps, without necessarily providing synthesis. However, unless we understand what happens when we intuit, we cannot be sure that there was no synthesis involved in the process of bypassing the steps of the analysis. It is also possible that bypassing the analytical steps happens by synthesizing these steps. As we will show later, rationality cannot be contrasted to intuition; on the one hand, the opposite of rational would be irrational, on the other hand, "ratio" means mind, so anything that comes from the mind is rational. Furthermore, we also know about many different forms of rationality, and some of these, such as Simon's "bounded rationality" leave ample space for intuition. Within the dual process theories, intuition is sometimes labeled parallel in contrast with the sequential mode of reasoning but what we know is that intuition is nonsequential; we are unsure whether it is parallel or not. Daniel Kahneman's (2011) recent work on fast and slow thinking seems to be to the point, although there are other ways of fast thinking beyond intuition, such as guessing. Due to limited knowledge and research, it seems that we will only discuss intuitive versus nonintuitive reasoning.

We distinguish between two kinds of intuition, these can be conceptualized as *intuitive judgment* and *intuitive insight* (see detailed argument in Dörfler & Ackermann, 2012; Stierand & Dörfler, 2016). Intuitive judgment is what we primarily associate with decision-making, more precisely, with a step in the decision-making process, usually called choice or decision taking, which is concerned with the evaluation of the decision alternative(s). Intuitive insight is, conversely, the intuition of the creatives; it is getting us to a solution to an ill-structured problem. It is important that it is about *a solution* rather than *the solution*, as a variety of new solutions can be created. However, intuitive insight may also appear in the decision-making process, only it is not associated with the decision taking phase but rather with creation (often mistakenly referred to as generation) of decision alternatives. However, in what follows, we primarily focus on intuitive judgment.

We want to emphasize that we do not argue for an exclusive use of intuition. What we expect to see in decision-making is a cycle of intuitive and nonintuitive steps, as described by Bergson (1911, 1946). Decision makers usually follow nonintuitive reasoning as long as they can, that is, as long as there is more information to gather and more time to gather it. However, when time is pressing and information is scarce, decision takers use intuitive judgment. They need to get into a nonintuitive mode again, to develop an explanation that justifies their intuitive judgment. The process is similar in the case of creativity, for example, when creating decision alternatives (see more details in Dörfler & Eden, 2014). It is important that the nonintuitive explanation always follows a flash of intuition and, although it often provides a meaningful explanation or even justification for the intuitive outcome, it may or may not have anything to do with what happened in the process of intuiting. Recognizing the cycles of intuition and nonintuition gains further importance when we consider that, for a long time, it was assumed that intuitive and nonintuitive reasoning are on the opposite ends of a single dimension and consequently the same person could only be good at one of these. Recent research (Hodgkinson, Sadler-Smith, Sinclair, & Ashkanasy, 2009), however, suggests that these are two different dimensions. Thus, we argue that for good decision-making we need good intuition as well as good nonintuition. The intuitive and the nonintuitive minds are friends, not foes.

Expert Intuition

> [...] with talent and a great deal of involved experience, the beginner develops into an expert who intuitively sees what to do without recourse to rules nor to remembered cases.
>
> **Hubert Dreyfus**

Our interest in intuition stems from a workshop in which one of us was involved with the board of executives of a large telecom company. As the importance of knowledge came up, it became clear that the board members only considered "textbook"-type knowledge, so we drew a quick schematic diagram about positioning intuition as a separate knowledge type (Dörfler Baracskai, Velencei, & Ackermann, 2011). While they quickly understood intuition as condensed expertise (Weick, 1995, p. 88), it was difficult to explain that it is not simply about experience, as experience does not automatically convert to expertise. In the words of Klein and Weick (2000, p. 19):

> The only thing that the passage of time achieves is to move you closer to retirement or termination. Too often, we treat experience as a noun rather than as a verb, something to accumulate.

Hence, experience is indispensable but not enough to become an expert. What matters most is what we do with that experience—we need to learn from it to develop expertise. In line with Dane and Pratt (2009), we see expertise as a precursor to trustworthy intuition; this view in the literature is emphasized by terms, such as "intuition-as-expertise" (Sadler-Smith & Shefy, 2004), "intuitive expertise" (Kahneman & Klein, 2009), or "expertise-based intuition" (Salas, Rosen, & DiazGranados, 2010).

The notion of expertise can also explain much of the disagreement in the management and organization studies field about the usefulness of intuition. If we take a closer look at those studies regarding the failure of intuition, (including Bowers, Regehr, Balthazard, & Parker, 1990, p. 97; Schoemaker & Russo, 1993, p. 27; Trailer & Morgan, 2004), we will usually find experiments targeted at intuitions of novices. Remarkably, we did not find a single instance where this was not the case. These experiments are typically, but not exclusively, conducted with students; for instance, Trailer and Morgan (2004) observed that undergraduate business school students made poor intuitive judgments in the field of physics. But, why would business school students have intuition in physics? In contrast, those who have found empirical (usually not gained through experiments) evidence of intuition working well in their respective fields of interest (including Burke & Miller, 1999 in management; Hayashi, 2001 in leadership; Keren, 1987 in the game of bridge), typically focused on intuition at high levels of expertise. Due to the relatively small number of studies providing empirical and experimental evidence about intuition in management, our argument is not conclusive but we find it intuitively convincing.

The significance of expertise for intuition can also be approached from the opposite end, from the development of expertise. There are three key models, developed using different methodological approaches, that explain all levels of expertise. The first was originally put forward by Simon and his various collaborators (e.g., Chase & Simon, 1973a,b; Gobet & Simon, 1996a,b, 2000), using an experimental approach. Then, the Dreyfus brothers developed their model using phenomenology (e.g., Dreyfus, 2004; Dreyfus & Dreyfus, 1986; Kreisler & Dreyfus, 2005). Last, Dörfler, Baracskai, and Velencei (2009) presented a purely speculative model of expertise levels. While these models have been developed using different methods, they are complementary and have a common touchpoint in acknowledging that intuiting becomes the dominant mode of knowing at the highest level of expertise.

Currently, there seems to be considerable agreements that intuition works well *only* at a high level of expertise (cf. Hogarth, 2001; Kahneman & Klein, 2009; Prietula & Simon, 1989; Salas, Rosen, & DiazGranados, 2010). Daniel Kahneman, who pointed out numerous flaws of intuition, always deliberately focused on commonsense-level intuition and criticized those painting a positive picture of intuition for not emphasizing that it is only intuition at a high level of expertise that works well—the intuition of experts he also finds useful.

The Process of Intuiting

> Intuition is not something that is given. I've trained my intuition to accept as obvious shapes which were initially rejected as absurd [...]
>
> **Benoit Mandelbrot**

We are unsure what happens when we intuit, for "much current knowledge rests largely on researchers' speculative arguments and abstract theorizations" (Sadler-Smith, 2016). Little empirical research has been done to date, only a fraction of this is qualitative, and only a small part of that focuses on the subjective experience of intuiting. Therefore, what we present here is how we speculate about what happens when we intuit, using Polányi's (1962, 1966a,b, 1983) work as our starting point (see Dörfler & Ackermann, 2012).

Polányi's original argument is concerned with *tacit knowing* more generally, but it also works for intuiting. Let us consider exploring a room with a stick with our eyes closed. Initially, we would be paying attention to what we feel in our fingers, such as the vibrations of the stick, the angle under which it is inclined, changes of direction, etc. These feelings are the *particulars* in the process of tacit knowing, they take place on the near end of the stick (*proximal term*), and they belong to the realm of *subsidiary knowing*. If we continue exploring the room a little longer, we will soon start picturing the room at the far end of the stick (*distal term*). Thus, our attention turns from the particulars to the *whole*, to the picture of the room that is still forming in our mind, and this whole is what we focus on, thus we label it as *focal knowing*. The process of tacit knowing is then an *integration* process, in which the particulars are integrated into the whole and the particulars seem to be dissolved in this process. Initially, some particulars will likely belong to the tacit realm, others to the explicit, but the integration process is tacit, and as soon as the focal whole emerges, we lose awareness of the particulars. We cannot tell anymore about the feeling in our palm, we can only tell how we picture the room (Figure 1.1).

The above example was perhaps not what immediately comes to mind when thinking of intuition, but has the advantage of having the particulars and the whole on the two different ends of the stick and thus we find it useful to start with. However, recognizing a face or writing a poem can be described in a very similar way. We could tell some characteristics of a face we recognize, before the actual recognition happens, and we can know the rules of grammar, letters, and so on, when writing a poem. However, when we are in the process of recognizing the face or writing the poem, we will have no idea which facial characteristics we have seen

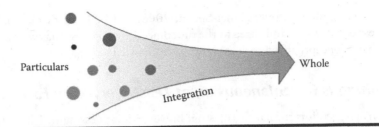

Figure 1.1 **The process of intuiting. (Adapted from Dörfler, V., & Ackermann, F.,** *Management Learning, 43*(5), 545–564, 2012.)

when the recognition took place or which rules of grammar we have used when the poem was written. What is interesting is that if we describe the facial characteristics of someone we have not seen for many years, we will recognize the face even if the characteristics we described beforehand have changed; we may not even notice that those characteristics have changed. Hence, we frequently don't even know what particulars we use to recognize the focal whole.

Although some would question whether recognizing a face or writing a poem qualifies for intuition, the process of intuiting works the same way in the case of intuitive judgments as well. The *particulars* would include our explicit expectations and the information we have about the decision alternatives but can also include things we may have no idea about, such as a move or a look of our negotiating partner. When we make the choice, we will not know what *particulars* we used. If we come up with an ex post explanation, that may or may not have anything to do with how we made that choice.

We are confident that the above description of the process of intuiting holds, but we could not say that it is particularly detailed. Although further empirical studies, particularly firsthand accounts of intuitions, may shed further light on some details, we must expect that we may never have a good and detailed description of the process of intuiting. However, good descriptions or models do not make for good intuition. The expertise of the person does.

Features of Intuition

Intuition is what you know for sure without knowing for certain.

Weston Agor

What makes distinguishing intuitive from nonintuitive reasoning so difficult is that there is no single characteristic along which this contrast can be made. In this section, we provide a set of six characteristics. These together can do the job; all six are necessary for identifying intuition, and if any one of them is missing, then it is something else occurring (cf. Dane & Pratt, 2007; Dörfler & Ackermann, 2012;

Kahneman, 2003, p. 698; Sadler-Smith, 2008, p. 13). Three of these refer to the process of intuiting, and three to the outcome of this process that we call intuition and have conceptualized earlier as intuitive knowledge.

Intuiting Is Instantaneous or, at Least, Very-Very Fast

In this respect, it is similar to guessing, however, it is guessing which is "frequently correct" (in line with Simon, 1983, p. 25). This speed of intuiting is particularly important today, when the time pressure is constant. As Handy (1995, p. 49) puts it, "[b]y the time you know where you ought to go, it's too late to go there." But how is intuiting so rapid? Per Prietula and Simon (1989, pp. 121–122), intuiting is a leap by which the expert bypasses the analytical steps and overcomes limitations of attention and of memory (both short-term memory [STM] and long-term memory [LTM]). We become aware of the right answer before consciously realizing it and without having to analyze everything simply by relying on our experience (Klein & Weick, 2000). This pattern recognition that helps bypass the steps of deliberate reasoning are not limited to situations the decision maker already knows. Expert decision makers will recognize patterns in new situations, not only in situations with which they are already familiar. This means that we should understand the patterns as *particulars* that are integrated in the whole in the process of intuiting, and this whole may be something we see for the first time.

Intuiting Is Spontaneous

Spontaneity means that intuiting does not require effort, at least at the moment when it happens (cf. Agor, 1984, p. 75). However, this also means that intuition cannot be produced at will (Isaack, 1978, p. 918). This does not mean that the decision makers just need to lay back and wait for intuition to arrive; hard work is needed beforehand, and then the intuiting happens in this relaxed state (for numerous examples, see Hadamard, 1954). The work needed for good intuition is not limited to the work on a problem; it includes all the previous work in the discipline or in the problem area and has a strong link to the level of expertise (Prietula & Simon, 1989). This means that intuiting brings together everything we have experienced in our field in various contexts and can have bearing on the decision situation at hand (cf. Rowan, 1986, p. 83). In terms of the intuiting process, we could say that the person can only integrate the particulars they have, and that involves all their knowledge in the discipline, in the problem area and about the context.

Intuiting Is Alogical

The terminology to describe this feature of intuiting is slightly different in every case: Kahneman and Tversky (1982, p. 124) describe it as "an informal and unstructured mode of reasoning, without the use of analytic methods or deliberate calculation," Barnard (1938, p. 301 ff) calls it a "nonlogical" process to contrast it to the logical process of reasoning, etc. However, the message is the same every time; intuiting operates *independently* of the general principles of reasoning that Russell (1946, p. 379) calls logic. We call this mode of reasoning *alogical*, meaning that it neither follows (*logical*) nor contradicts (*illogical*) the rules of logic. For similar reasons, we can describe intuiting as arational, meaning that it is not rational or irrational, simply independent from the rules of rationality. It is important to note that qualifying intuiting as alogical does not tell us anything about its modus operandi, we only know what it is not. The question is whether we will ever discover some set of rules that intuition seems to follow. We believe that it is likely that we will not, as the *particulars* are of very personal natures, embracing apart from the accepted knowledge in the discipline and in the problem area also the subjective aspects of these as well as of the current context and also the personal history of the person. However, we agree with Simon (1987, p. 61) that "... intuition is not a process that operates independently of analysis; rather, the two processes are essential complementary components of effective decision making systems."

Intuition Is Gestalt or a Holistic Hunch

Intuition is gestalt or a holistic hunch as mentioned by Beveridge et al. (Beveridge, 1957, p. 73; Hayashi, 2001, p. 64; Miller & Ireland, 2005; Morris, 1967, p. 158; Sinclair & Ashkanasy, 2005, p. 357). This insight means two things. On the one hand, intuition is about the "big picture" of the decision situation, including the broad context, far reaching implications (and the implications of the implications, etc.) that are usually not considered in step-by-step reasoning, as the probabilities would be considered very low. The decision situation also considers what we could call "invisible parts" that are inaccessible to deliberate step-by-step reasoning, which are sometimes referred to as the *unknown unknowns*. On the other hand, intuition also involves the totality of the person. This view is also fully in harmony with how the Dreyfus brothers (see "Expert Intuition" section) described the highest level of expertise, that is, that the totality of the situation is intuitively perceived and the response involves the complete personality in an intuitive response. This does not mean, however, that intuition needs to be vague. Intuition is about the "essence;" it means seeing the "big picture" as well as the relevant details, and being able to quickly switch between the two (Dörfler & Eden, 2014). Thus, the (expert) person will be able to see which detail needs to be changed, as well as how this will affect the big picture.

Intuition Is Tacit

Hayashi (2001, p. 60) asserts, based on numerous interviews, that top executives cannot describe the process of intuiting much beyond labeling it "professional judgment," "intuition," "gut instinct," "inner voice," or "a hunch." Dane and Pratt

(2007, p. 36) therefore characterize intuition as nonconscious, meaning that the outcomes of intuiting are accessible to conscious thinking, but how one arrives at them is not. On the one hand, this seems sensible based on how we described the process of intuiting in the previous section. On the other hand, tacitness is very difficult to accept for the schooled mind, as our schools educate us to require sound reasoning or even proof for our choices. However, this is only as we are focusing on the wrong side of the schooling when we think about decision-making, on the mathematical mind. In the arts, if we paint a picture or write a poem, we are not required to provide a justification for why we see things in a certain way. So, we are back to Barnard's point of talking about the "executive arts." If we want to make use of our intuition, we must accept that it is tacit. This, however, does not mean that we cannot provide a nonintuitive justification that explains why the intuitive judgment makes sense, only we need to know that this explanation is likely, not how we arrived at the intuitive judgment—and that we will never know how we arrived at it. Thus, we can say that the tacit nature of intuition is an additional reason for why we need intuition and nonintuition to work in cycles.

Intuitors Are Confident about Their Intuitions

The explanations discussed under the previous section serve to make others accept the intuitor's judgment—the intuitors usually do not need justification. Jung (1921, § 770) distinguished four psychological functions: thinking, feeling, sensation, and intuition. He emphasizes that intuitive knowledge "possesses an intrinsic certainty and conviction." This certainty is one of the most commonly described features of intuition emphasized by both decision makers and great scientists. One of the most often quoted examples is Poincaré's story (e.g., Damasio, 1994; Goldberg, 1983b; Hadamard, 1954; Polányi & Prosch, 1977; Vaughan, 1979); likely because it is striking that his intuition "told him" the opposite of what he was trying to prove previously and later he proved this opposite. It is important to note that the built-in feeling of certainty of intuitive knowledge does not mean that intuition is infallible. Sadler-Smith (2008, p. 28) quotes critiques of intuition saying that "intuition is sometimes wrong but never in doubt." No one claims that nonintuitive reasoning is always correct. We also must not expect this from intuition. However, when it is the nonintuitive reasoning that fails us, this is easier to accept, as we simply did some steps that can be checked afterwards, and it is relatively easy to blame the failure to something external. However, if intuition is tacit and we don't have much to go on than saying that "I feel confident that I am right" and then we find that we were wrong, we can only blame ourselves. Not to mention, others will blame us as well. However, if we also know that intuition is not infallible, we have a good starting point for our intuitive endeavors.

There are two further things that are mentioned as features of intuition/intuiting. The first are the somatic effects, the second are the affective charges. Bodily feelings (somatic effects) are often associated with intuition, as this seems to be part of the language of how our intuitions speak to us. Expressions such as "gut feeling," "butterflies in the stomach," and similar are not simply metaphors; many intuitors have their somatic effects tell them that they have arrived at an intuitive judgment. We do not consider the somatic effects as features of intuition, as there seem to be a great variety, and they do not seem to be a necessary aspect of intuition and nonintuitive judgments may also be accompanied with somatic effects. We can also hear that intuitions are accompanied with emotional effects. Decision makers often seem to be married to their intuitive judgments. Most authors, however, are very cautious about the affective charges, and usually we can read that intuitions are "often affectively charged," suggesting that this may not always be the case. We will probably learn more about this, considering that intuition research is still in its infancy. However, we have seen decision makers married to their statistics tables, feasibility studies, and pie charts. Therefore, we do not consider emotions to be a necessary feature of intuition, but if we notice emotions we are more likely to expect intuition.

Enhancing Intuition

> [...] the methods of scientific inquiry cannot be explicitly formulated and hence can be transmitted only in the same ways as an art, by the affiliation of apprentices to a master.
>
> **Michael Polányi**

From what we have said so far about intuition, we believe there is no doubt about the importance of intuition for expert decision makers. However, based on the process and features of intuition, it does not seem straightforward that one can become better at intuiting. We have shown that at a high level of expertise intuition becomes the dominant mode of knowing. Similarly, we have shown that reliable intuition is only available at high level of expertise. Therefore, it is probably clear that becoming more knowledgeable in a particular knowledge domain helps in bettering intuition. We could consider this to be a generic way of enhancing intuition. The second group would include direct ways of enhancing intuition, meaning that these ways would focus on directly making intuition better. These direct ways often involve various mental exercises that the intuitors can perform themselves or in groups with or without supervision. There is, however, one direct way of enhancing intuition that stands out: the master-apprentice relationship. The third group of intuition enhancement ways we call indirect ways, as their focus is not on intuition itself, but something that relates to or interacts with intuition, namely the context of the intuitor, the awareness of intuition and the action taken based on intuition. In this section, we do not engage in the generic way, as that could include everything that we know about learning. We also do not tackle most of the direct ways, as the academic basis of these are not particularly elaborate. The exception is the master-apprentice relationship, which we briefly describe, as it stands out both in its significance as well as how much we know about it. We also provide brief description of the indirect ways.

The Master-Apprentice Relationship

The master-apprentice relationship is a widely accepted mode of achieving the highest level of expertise, and we have explored it at length elsewhere. (Baracskai, Dörfler, & Velencei, 2005; Dörfler & Eden, 2014; Dörfler & Stierand, 2009; Stierand, 2015) We have argued that it seems to be the only known way of passing on tacit knowledge, and thus the only direct way of learning intuition from someone else. The essence of the master-apprentice relationship is that it is wrong to follow and to abandon the master's way; however, from this struggle, a new master emerges. What happens in regarding intuition is best expressed by Polányi (1966a, p. 14): "A novice, trying to understand the skill of a master, will seek mentally to combine his movements to the pattern to which the master combines them practically." So, the apprentice is watching the master using her/his intuition, perhaps asking questions about it, and then will apply what Polányi (1946) termed "intelligent imitation"—which is imitation that is adapted to personality, context, and problem. Of course, the master provides feedback on the apprentice's intuition and the apprentice gradually evolves into a master.

Creating an Intuition-Friendly Environment

It is often emphasized in the literature, as well as in personal communication with intuitive people, that they need to hide the intuitive origins of their achievements and provide a post-rationalized well-structured argument instead. Such environments discourage the use of intuition. Therefore, creating intuition-friendly environments at personal, interpersonal and organizational levels can make a great deal of difference. At a personal level, this means putting our minds in a state where they are more open to intuition and accepting intuition. For example, it is well documented that reading poetry; enjoying art, music, or extreme sports; sitting in the woods or consuming "philosophical food" (Agor, 1984, p. 75) can help liberate the mind from "uninspiring" problems and may foster intuitive and creative ways of thinking. The interpersonal context means people who accept intuition as a valid form of knowledge can discuss it with the intuitor. Sadler-Smith and Shefy (2007), for instance, emphasize the importance of promoting the development of intuition through feedback. Finally, in an intuition-friendly organizational context it is acceptable to admit to using intuition, which does not mean that an argument does not need to be provided to substantiate the intuitive judgment. However, those who proved themselves as good intuitors may gradually enjoy benefits of providing the argument later or not at all. This is particularly important for organizations in a turbulent environment, when the response time is of crucial importance.

Increasing Awareness of Intuition

Experienced intuitors are usually good at recognizing and interpreting their intuitions. However, those who just rely on their intuitions often fail to notice or find it difficult to distinguish intuition from other phenomena, including hopes and fears, that may have similar somatic and affective characteristics. Knowing our bodies and emotions is very important in this sense. Personal level self-observation, reflection and keeping diaries can help to know your body and mind (Cartwright, 2004; Goldberg, 1983a, pp. 193–194; Vaughan, 1979, p. 205). At an interpersonal level discussing intuitive experiences with peers and persons of trust can be helpful (e.g., Agor, 1984, p. 66). At an organizational level, we can only be supportive of the use as well as development of intuition; not only in terms of the above noted techniques, but also by supporting master-apprentice relationships.

Acting upon Intuition

Creating a supportive environment is likely to help intuition occur, better awareness will help to notice and understand intuitive leaps, but then we also need to act upon intuition. When the intuition happens, a person should not delay the action. We should think it through again, check by means of structured step-by-step reasoning "just in case" and so forth. If intuition is not followed by action, we will never know whether it has worked in the first place. There are no techniques suggested explicitly for improving *how* we can act upon intuition but many of the previously mentioned techniques also support action. Reflecting helps learning from previous actions and poetry, art, and philosophy may help to achieve an actionable mindset. Discussing it with peers provides examples as well as feedback. An organizational environment that supports intuition also supports acting upon it; most importantly, it does not sanction severely when intuition led to bad outcome through action. Of course, intuition is not always right, and neither is the nonintuitive step-by-step reasoning. While we cannot allow poor outcomes all the time, in the case of nonintuitive step-by-step reasoning people are not usually penalized for a single failure—intuitors should be treated the same.

While we cannot simply set up courses or read a couple of books to increase our intuition, we wanted to show that there is much that we can do to support the development of intuition. Several of these things are not unique to intuition and not so alien to our culture, such as peer discussions, reflection and reflexivity, the master-apprentice relationship, etc. We make huge efforts to improve our nonintuitive step-by-step reasoning capabilities—if we do as much for intuition, it should be sufficient.

Is Intuition Mystical?

I propose that the goal of science is to make the wonderful and the complex understandable and simple—but not less wonderful.

Herbert Simon

Having gone through the process of polishing the initial picture about intuition from various perspectives, we return to the initial question: Is intuition mystical? Lieberman (2000, p. 109) suggests that intuition is at best regarded as mysterious and unexplainable. To the contrary, Davenport and Prusak (2000, p. 11) argue: "We arrive at an answer intuitively, without knowing how we got there. That does not mean the steps not to exist—intuition is not mystical. It means we have thoroughly learned the steps that they happen automatically, without conscious thought, and therefore at great speed."

Our answer is somewhere in-between these two. We believe that it is useful to try to understand intuition in the academic sense of the word. However, as intuiting happens almost instantaneously and intuition is tacit, there is a limit to this. We agree with Isaack (1978, p. 919) that "intellect cannot completely understand the intuition since the artificial tools, preconceived categories, and symbols used by the intellect only represent reality and are not the substance of reality." However, this should not stop us from trying to understand intuition. We need intuition to make sense of intuition.

References

Agor, W. H. (1984). *Intuitive management: Integrating left and right brain management skills.* Englewood Cliffs, NJ: Prentice Hall.

Agor, W. H. (1986). The logic of intuition: How top executives make important decisions. *Organizational Dynamics, 14*(3), 5–18. doi:10.1016/0090-2616(86)90028-8

Baracskai, Z., Dörfler, V., & Velencei, J. (2005). *Majstor i kalfa* [Master and apprentice]. Croatia: Sinergija.

Barnard, C. I. (1938/1968). *The functions of the executive.* Cambridge, MA: Harvard University Press.

Bergson, H. (1911). *Creative evolution.* New York, NY: Henry Holt and Company.

Bergson, H. (1946/1992). *The creative mind: An introduction to metaphysics.* New York, NY: Citadel Press.

Beveridge, W. I. B. (1957/2004). *The art of scientific investigation* (2nd ed.). Caldwell, NJ: Blackburn Press.

Bowers, K. S., Regehr, G., Balthazard, C., & Parker, K. (1990). Intuition in the context of discovery. *Cognitive Psychology, 22*(1), 72–110. Retrieved from http://www.sciencedirect.com/science/article/B6WCR-4D5XC4T-3G/1/eef8eb477b34a852db158e2f37f9be79

Burke, L. A., & Miller, M. K. (1999). Taking the mystery out of intuitive decision making. *Academy of Management Executive, 13*(4): 91–99. Retrieved from http://dx.doi.org/10.5465/AME.1999.2570557.

Cartwright, T. (2004). Feeling your way: Enhancing leadership through intuition. *Leadership in Action, 24*(2), 8–11. Retrieved from http://proquest.umi.com/pqdweb?did=670515 911&sid=1&Fmt=2&clientId=46002&RQT=309&VName=PQD.

Chalmers, D. J. (1998). *The conscious mind: In search of a fundamental theory* (paperback ed.). New York, NY: Oxford University Press.

Chase, W. G., & Simon, H. A. (1973a). The mind's eye in chess. In W. G. Chase (Ed.), *Visual information processing* (pp. 215–281). New York, NY: Academic Press.

Chase, W. G., & Simon, H.A. (1973b). Perception in chess. *Cognitive Psychology, 4*(1), 55–81. Retrieved from http://www.sciencedirect.com/science/article/B6WCR-4D6RJR4-53/2/d368b90478edf5d8b9ee81a894eff330.

Damasio, A. R. (1994/2005). *Descartes' error: Emotion, reason, and the human brain.* New York, NY: Penguin Books.

Dane, E., & Pratt, M. G. (2007). Exploring intuition and its role in managerial decision making. *Academy of Management Review, 32*(1), 33–54. doi:10.5465/AMR.2007.23463682.

Dane, E., & Pratt, M. G. (2009). Conceptualizing and measuring intuition: A review of recent trends. In G. P. Hodgkinson & J. K. Ford (Eds.), *International review of industrial and organizational psychology* (Vol. 24, pp. 1–40). Chichester: Wiley. Retrieved from http://media.wiley.com/product_data/excerpt/08/04706800/0470680008.pdf

Davenport, T. H., & Prusak, L. (2000). *Working knowledge: How organizations manage what they know* (paperback ed.). Boston, MA: Harvard Business School Press.

Dörfler, V., & Ackermann, F. (2012). Understanding intuition: The case for two forms of intuition. *Management Learning, 43*(5), 545–564. doi:10.1177/1350507611434686.

Dörfler, V., Baracskai, Z., & Velencei, J. (2009, August 7–11). *Knowledge levels: 3-D model of the levels of expertise.* AoM 2009: The 69th Annual Meeting of the Academy of Management, Chicago, IL.

Dörfler, V., Baracskai, Z., Velencei, J., & Ackermann, F. (2011). *Facts, skills and intuition: A typology of personal knowledge.* Management science working papers, Strathclyde University, Glasgow.

Dörfler, V., & Eden, C. (2014, August 1–5). *Understanding "expert" scientists: Implications for management and organization research.* AoM 2014: The 74th Annual Meeting of the Academy of Management, Philadelphia, PA. Electronic version: 10.5465/AMBPP.2014.10732abstract.

Dörfler, V., & Stierand, M. (2009, September 15–17). *Investigating the extraordinary.* Brighton: BAM.

Dörfler, V., & Szendrey, J. (2008, April 28–30). *From knowledge management to cognition management: A multi-potential view of cognition.* International Conference on Organizational Learning, Knowledge and Capabilities (OLKC), Copenhagen, Denmark. Electronic version: http://www.viktordorfler.com/webdav/papers/MultipotentialCognition.pdf.

Dreyfus, H. L. (2004). A Phenomenology of Skill Acquisition as the basis for a Merleau-Pontian Nonrepresentationalist Cognitive Science (unpublished). Electronic version: http://ist-socrates.berkeley.edu/~hdreyfus/pdf/MerleauPontySkillCogSci.pdf.

Dreyfus, H. L., & Dreyfus, S. E. (1986/2000). *Mind over machine.* New York, NY: The Free Press.

Gobet, F., & Simon, H. A. (1996a). Recall of random and distorted chess positions: Implications for the theory of expertise. *Memory & Cognition, 24*(4), 493–503. Retrieved from http://www.psychonomic.org/search/view.cgi?id=1334.

Gobet, F., & Simon, H. A. (1996b). Templates in chess memory: Mechanism for re-calling several boards. *Cognitive Psychology, 31*(1), 1–40. doi: 10.1006/cogp.1996.0011.

Gobet, F., & Simon, H. A. (2000). Five seconds or sixty? Presentation time in expert memory. *Cognitive Science, 24*(4), 651–682. Retrieved from http://www.sciencedirect. com/science/article/B6W48-41WSC68-4/2/83c8cd00262a31865c778304223942c3.

Goldberg, P. (1983a). The intuitive experience. In W. H. Agor (Ed.), *Intuition in organizations: Leading and managing productively* (pp. 173–194). Newbury Park, CA: Sage Publications.

Goldberg, P. (1983b). The many faces of intuition. In W. H. Agor (Ed.), *Intuition in organizations: Leading and managing productively* (pp. 62–77). Newbury Park, CA: Sage Publications.

Hadamard, J. (1954). *The psychology of invention in the mathematical field.* New York, NY: Dover Publications.

Handy, C. (1995). *The empty raincoat: Making sense of the future.* London: Arrow Business Books.

Hayashi, A. M. (2001). When to trust your gut. *Harvard Business Review, 79*(2), 59–65.

Hodgkinson, G. P., Sadler-Smith, E, Sinclair, M., & Ashkanasy, N. M. (2009). More than meets the eye? Intuition and analysis revisited. *Personality and Individual Differences, 47*(4), 342–346. Retrieved from http://www.sciencedirect.com/science/article/ B6V9F-4W84GR5-2/2/a60ac4b2b44ee7b34d4a4b9ac7f52498.

Hogarth, R. M. (2001). *Educating intuition.* Chicago, IL: University of Chicago Press.

Isaack, T. S. (1978). Intuition: An ignored dimension of management. *Academy of Management Review, 3*(4), 917–922. doi:10.5465/AMR.1978.4289310.

Jung, C. G. (1921/1990). *Psychological types.* Princeton, NJ: Princeton University Press.

Kahneman, D. (2003). A perspective on judgment and choice: Mapping bounded rationality. *American Psychologist, 58*(9), 697–720. Retrieved from http://content2.apa.org/ journals/amp/58/9/697.

Kahneman, D. (2011). *Thinking, fast and slow.* London: Penguin Books.

Kahneman, D., & Klein, G. (2009). Conditions for intuitive expertise: A failure to disagree. *American Psychologist, 64*(6), 515–526. Retrieved from http://psycnet.apa.org.proxy. lib.strath.ac.uk/journals/amp/64/6/515.html.

Kahneman, D., & Tversky, A. (1982). On the study of statistical intuitions. *Cognition, 11*(2), 123–141. doi:10.1016/0010-0277(82)90022-1.

Keren, G. (1987). Facing uncertainty in the game of bridge: A calibration study. *Organizational Behavior and Human Decision Processes, 39*(1), 98–114. Retrieved from http://www. sciencedirect.com/science/article/B6WP2-4CYG2X0-BY/2/47bc4960fae485620d80e fe2f7b98efe.

Klein, G., & Weick, K. E. (2000). Decisions: Making the right ones. Learning from the wrong ones. *Across the Board, 37*(6), 16–22. Retrieved from http://search.ebscohost. com/login.aspx?direct=true&db=ofs&AN=510099826&authtype=shib&site=eh ost-live.

Kreisler, H., & Dreyfus, H. L. (2005). *Meaning, relevance, and the limits of technology,* Interview with Hubert Dreyfus (Conversations with History), Berkeley, CA: Institute of International Studies, University of California.

Lieberman, M. D. (2000). Intuition: A social cognitive neuroscience approach. *Psychological Bulletin, 126*(1), 109–137. Retrieved from http://content2.apa.org/journals/ bul/126/1/109.

Miller, C. C., & Ireland, R. D. (2005). Intuition in strategic decision making: Friend or foe in the fast-paced 21st century?. *Academy of Management Executive, 19*(1), 19–30. doi:10.5465/AME.2005.15841948.

Mintzberg, H. (1994). The fall and rise of strategic planning. *Harvard Business Review, 72*(1), 107–114.

Morris, W. T. (1967). Intuition and relevance. *Management Science, 14*(4), 157–165. doi:10.1287/mnsc.14.4.B157.

Polányi, M. (1946). *Science, faith and society.* London: Oxford University Press.

Polányi, M. (1962/2002). *Personal knowledge: Towards a post-critical philosophy.* London: Routledge.

Polányi, M. (1966a). The logic of tacit inference. *Philosophy, 41*(155), 1–18. doi:10.1017/S0031819100066110.

Polányi, M. (1966b/1983). *The tacit dimension.* Gloucester, MA: Peter Smith.

Polányi, M., & Prosch, H. (1977). *Meaning* (paperback ed.). Chicago, IL: The University of Chicago Press.

Prietula, M. J., & Simon, H. A. (1989). The experts in your midst. *Harvard Business Review, 67*(1), 120–124.

Rowan, R. (1986). What it is. In W. H. Agor (Ed.), *Intuition in organizations: Leading and managing productively* (pp. 78–88). Newbury Park, CA: Sage Publications.

Russell, B. A. (1946/2004). *History of western philosophy,* London: Routledge.

Sadler-Smith, E. (2008). *Inside intuition.* London, UK: Routledge.

Sadler-Smith, E. (2016). 'What happens when you intuit?': Understanding human resource practitioners' subjective experience of intuition through a novel linguistic method. *Human Relations, 69*(5), 1069–1093. doi:10.1177/0018726715602047.

Sadler-Smith, E., & Shefy, E. (2004). The intuitive executive: Understanding and applying 'gut feel' in decision-making. *Academy of Management Executive, 18*(4), 76–91. doi:10.5465/AME.2004.15268692.

Sadler-Smith, E., & Shefy, E. (2007). Developing intuitive awareness in management education. *Academy of Management Learning & Education, 6*(2), 186–205. doi:10.5465/AMLE.2007.25223458.

Salas, E, Rosen, M. A., & DiazGranados, D. (2010). Expertise-based intuition and decision making in organizations. *Journal of Management, 36*(4), 941–973. Retrieved from http://vnweb.hwwilsonweb.com/hww/jumpstart.jhtml?recid=0bc05f7a67b1790 e029d610c70c618e1e7a103d48ec260a169ce81520d132cdaefb009c97c7503ee& fmt=H.

Schoemaker, P. J. H., & Russo, J. E. (1993). Pyramid of decision approaches. *California Management Review, 36*(1), 9–31. Retrieved from http://proquest.umi.com/pqdweb?d id=288936&sid=1&Fmt=3&clientId=46002&RQT=309&VName=PQD.

Simon, H. A. (1947). *Administrative behavior: A study of decision-making processes in administrative organization* (1st ed.). New York, NY: Macmillan.

Simon, H. A. (1983). *Reason in human affairs.* Stanford, CA: Stanford University Press.

Simon, H. A. (1987). Making management decisions: The role of intuition and emotion. *Academy of Management Executive, 1*(1), 57–64. doi:10.5465/AME.1987.4275905.

Sinclair, M., & Ashkanasy, N. M. (2005). Intuition: Myth or a decision-making tool?. *Management Learning, 36*(3), 353–370. doi:10.1177/1350507605055351.

Spinoza, B. (1677/2000). *Ethics: Demonstrated in geometrical order.* New York, NY: Oxford University Press.

Stierand, M. (2015). Developing creativity in practice: Explorations with world-renowned chefs. *Management Learning, 46*(5), 598–617. doi:10.1177/1350507614560302.

Stierand, M., & Dörfler, V. (2016). The role of intuition in the creative process of expert chefs. *Journal of Creative Behavior, 50*(3), 178–185. doi:10.1002/jocb.100.

Trailer, J. W., & Morgan, J. F. (2004). Making "good" decisions: What intuitive physics reveals about the failure of intuition. *Journal of American Academy of Business, Cambridge,* 4(1–2), 42–48. Retrieved from http://proquest.umi.com/pqdlink?did=52 4073511&sid=4&Fmt=4&clientId=46002&RQT=309&VName=PQD.

Vaughan, F. E. (1979). *Awakening intuition.* New York, NY: Anchor Books.

Weick, K. E. (1995). *Sensemaking in organizations.* Thousand Oaks, CA: Sage Publications.

Chapter 2

Intuition as a Complement to Analytics

Lionel Page

Contents

Human behavior flows from three main sources: desire, emotion, and knowledge.

Plato

Introduction

Freud stated that his work on the subconscious was a third outrage to humanity's self-esteem (Freud & Strachey, 1977). First, Copernicus had shown that humans were not located at the center of the universe. Then, Darwin had shown that humans had not been specially created outside of the animal world. Finally, Freud showed that humans were not masters in their own mind. While there is not much scientific content to take away from Freud's work, this insight is still relevant in some sense to understand the progress of our understanding of the human mind. This progress has followed a dialectical path with opposite views replacing each other over time.

Historically, the strength of the human mind has long been associated with its ability to engage in rational thinking, the conscious use of abstract reasoning to solve. Intuitions, emotions and subconscious processes were viewed as dangers to our rational minds. The rational action theory, which was, until recently, at the core of the economic science, was in perfect harmony with such a vision. Since its inception in the early twentieth century, this theory was purposely built without psychology content. However, in the second part of the century, psychologists led a scientific offensive against this paradigm, showing all the psychological flaws of the human mind, grounded in subconscious processes, emotions, and automatic responses. The "heuristic and biases" paradigm accumulated evidence about failures of human decision-making processes. To the perfect rational actor from economics, it substituted a worrying picture of an individual without consistent motives and without coherent decisions processes to achieve his goals. In some ways, this paradigm was pushing a humbling view of human consciousness echoing Freud's remark. Unlike the naïve optimism of the rational action model, psychologists were suggesting that humans are quite hopeless at what they are doing, and that they are oblivious to this hopelessness.

As often in science, progress leads to going beyond oppositions between different views. New evidence from economics, psychology, and neuroscience point to a vision of the human mind which takes both from the rational action theory and from the role of subconscious processes. In this chapter, I will argue that the present evidence suggests seeing intuitions/subconscious processes and conscious rationality as complementing aspects of our cognition, designed to help us make good decisions in a complex environment. The question for the modern decision maker is not how to rely purely on intuitions/feelings or on analytical rationality, it is how to tap into these two elements to make the best decisions.

From Cold Rationality to Human Imperfection

Economics is at the heart of behavioral science. From its etymology *oikonomos*, the management of the household, economics is the science of decisions to allocate resources to competing uses—like consumption and production. Arguably, economics is one of the most formal and unified disciplines of the behavioral sciences (Gintis, 2007). When compared to psychology or sociology, economics is characterized by a unified conceptual framework based on shared assumptions upon which formal models are built.

The formalism of the economic approach sets it apart from other behavioral disciplines. Perhaps the formal and quantitative nature of economics originates in the fact that economic is about making the best decisions with measurable quantities (e.g., working time, money, goods). In any case, the discipline followed a very distinct path to other behavioral sciences. From the beginning of twentieth century onwards, economics expelled psychology from its core theories to develop a study of economic decision based on rational decision makers (Bruni & Sugden, 2007).

The idea was very much in tune with the positivist argument that science should only focus on what is observable (Hahn, Neurath, & Carnap, 1996). The human psyche was not observable and should therefore be left outside of the realm of the scientific inquiry about economic decisions. Economists would look at people's choices. The empirical study of what drives people's preferences was left to other disciplines such as psychology. Economists instead opted to make minimal behavioral assumptions considered so reasonable that they could be called axioms (i.e., evidently true). Specifically, economists "only" assumed that people have preferences over goods (they know what they want) and that whatever these preferences, they are consistent in some simple ways (not self-contradictory).

From that starting point, formal mathematical methods could be used to determine what would be the best decisions for an agent given his preferences. Whether this approach was normative (describing what agents should do) or positive (describing what agents do) was not always clear. The models unveiled the best decisions of rational agents, so surely they were normative in the sense that deviating from these models' predictions would be worse in some sense. But models also became considered as ways of describing actual behavior, partly due to the fact that it was the only way economists knew to investigate behavior.

The lines between the positive and normative nature of economic models of behavior were blurry. Many economists had a pragmatic approach, saying that models had the merit to exist and that a model is better than no model. Surely, human behavior was not as perfect as modeled but if deviations were minor, models offered a good approximation. At the same time, many economists adopted a more positive approach. Models were based on reasonable principles of rationality which would have been strange for people not to respect. Hence, economics was describing the actual behavior of decision makers (Stigler & Becker, 1977).

To those outside of economics, the decision maker described by economics textbooks was however strange and strikingly different from the common experience. It was a person who knew what he/she wanted without doubting about his/her preferences, could plan well ahead to reach his/her goals, could interpret complex signals accurately to form accurate beliefs, and could engage in complex computations to find out optimal solutions to his/her problems. This person was the perfect image of cold rationality.

In the second part of the twentieth century, a growing amount of work by psychologists and experimental economists pointed to a range of problems with this model of human behavior. Placed in controlled behavioral experiments, humans do not excel at computations; they often misinterpret noisy signals leading to distorted beliefs, and they can fail to stick to their plans leading to inconsistent choices over time. Even more worrying, it appeared that humans do not respect one of the basic principles of rationality, the coherence of preferences, (Kahneman, 2012).

This work, led by psychologists such as Kahneman and Tversky, eventually formed the foundations of the behavioral revolution in economics. A huge amount of empirical evidence was accumulated showing a radically different view of humans as imperfect and biased decision makers. While homo *economicus* was the image of a perfect rational decision maker, homo behavioralis was a seriously impaired one. His preferences were often incoherent or just made up on the fly, his beliefs were distorted in many systematic ways, and his decision processes were slow and prone to errors.

However, paradoxically, this part of the behavioral revolution had kept something from the old economics textbooks. While it had rejected homo *economicus* as a positive theory describing how people make decisions in the real world, it kept it as a normative theory of how people should behave if they want to make good decisions. In this behavioral approach, deviations from the standard model are "biases," which lead people to make costly mistakes. The ability of humans to reasons (so-called system 2) must be used to correct mistakes from our intuitions (coming from the so-called system 1) (Kahneman, 2003, "A Perspective on Judgment and Choice: Mapping Bounded Reality").

In some way, at the heart of this approach, the traditional figure of cold rationality as an ideal is still present; an ideal which is not reached by real humans. In that view, decision makers need supporting tools to help make their decisions more rational/optimal, less influenced by their flawed psychology.

Heuristics as Adaptive But Cold Rationality Not Useful

In addition to the "heuristics and biases" approach from Kahneman and Tversky, another one in psychology proposed a radically different view about human decision-making. The psychologist Gerd Gigerenzer argued that heuristics, the psychological rules of thumb human use in their everyday life (instead of the complex computations assumed by the homo economicus model), are simple rules that make us smart (Gigerenzer, 2000). They are cognitive shortcuts that allow us to find quick solutions that are good enough in many situations in the complex environment we live in.

A key argument at the heart of Gigerenzer's view is the idea that problems to solve in the real world are incredibly complex—so complex as to be de facto out of bounds for reasonable computational abilities within a limited time frame. From that point of view, the idea that decision makers could find the "optimal solution" is naïve. Many problems that economic decision makers face are "computationally hard," in that there is no algorithm to solve the problem whose computational time grows as a polynomial function of the size of the problem. Problems as mundane as selecting the best basket of goods in a supermarket for a given budget limit are computationally hard. While it may be easy to find a good enough solution, finding the optimal solution to such problems can require enormous computational effort.

To understand how decision makers live in such a complex world, the concept of a perfect Spock-like decision maker can be a misleading role model. Instead of idealizing cold rationality, one may want to understand how people make good decisions in the real world. As the outcome of eons of time of evolution, humans cannot be entirely flawed decision makers. It is more likely that the features of their decision-making processes are adaptive solutions to the problems they face.

A key aspect of this approach was not to dispel aspects of human psychology that do not fit with the model of cold rationality. Gigerenzer defended that things like emotions, intuitions and gut feelings should be appraised positively for the role they play in human decision-making. By providing ready-made answers, heuristics, in complex situations, can help humans reach ecological rationality: making good decisions in the context of the real world.

The influential research in neuroscience from Antonio Damasio has given further weight to such a view. It has shown that cold rationality is not the key to good decisions with emotions potentially leading us astray. Instead, emotions are the key to decision-making (Damasio, 2006). A huge amount of information about the surrounding world is processed automatically by brain processes producing positive/negative values as output to help making decisions. These emotions/intuitions are positive/negative feelings that are not brought to the fore of consciousness in many situations. They are, however, essential to allow us to make even the simplest decisions. Some patients with damage to the ventromedial prefrontal cortex can lose the coupling of bodily arousal with decisions (Bechara, Damasio, Damasio, & Anderson, 1994). They can maintain normal intellect and can make rational judgments, but become unable to make mundane choices such as picking a variety of potatoes in the supermarket. While humans with a fully functioning brain would quickly get a feeling for the best option and pick it often without thinking about it, these persons will not feel anything. While being able to describe all the characteristics of the options (e.g., shape, price) and the intended use (e.g., soup), they will feel blank about which option is best. Without emotions, they are literally stuck in the allegorical positions of Buridan's donkey who dies of hunger and thirst while being placed at equal distance of food and water.

Since Damasio's contribution, the progress in neuroscience has pointed to the large amount of computations made by the brain beyond consciousness (Montague, 2007). The key role of a brain is to make decisions, as only decisions have in the end an impact on the individual's fitness. Decisions require assessment of the different options available and their value. Behind most simple choices, the human brain is processing a large amount of new information to assess quickly the best option to pick.

The challenge is that the sheer quantity of information to process is too large to be extensively processed. Consider a simple walk in the supermarket. The brain must process rich visual information about the surrounding topography; the identity of a wide range of goods, and the presence of other human beings and moving objects (carts). It will also have to integrate this information with auditory information containing background noises, voices of other surrounding human beings, and background music or supermarket announcements. This information is nowhere near enough to make decisions. Just to decide where to go in the supermarket, the human brain must form a goal about what is needed in relation to stated goals (e.g., dinner). It must decide what path is the best to get there, and must interpret the intentions of other humans moving around from observing them, to avoid bumping into them. It has then to look at similar goods and assess which one is the most likely to fit the intended use for a low price. To appreciate the complexity of the amount of information a human brain must handle to solve such a task that is almost second nature to us, one can observe that despite having computers with tremendous computational power, we are still not able to design robots that carry mundane human activities.

Given the amount of information to process, a key ingredient is optimally allocating cognitive resources to solve the problem at hand. It is achieved by heavily relying on memory. From all of our experiences with the outside world, we have built an internal model of it which we do not need to update. Any new stimulus fitting this model do not have to be re-interpreted and only deviations from all our expectations need to be processed. In that way, our brain can focus on just a thin slice of reality to make decisions (Gladwell, 2007). Going back to our supermarket example, as we walk in the aisles we do not need to consider the information from all our visual stimuli to consider whether the objects pushed around by other shoppers are carts. The visual features of these objects are quickly associated with known visual patterns associated with carts, which are expected in this environment. Consequently, our visual information about these carts can be largely ignored for our attention to focus on the question for which we do not have an easy solution: what we should get for dinner.

In some ways, the voice in our head is an illusory representation of our self. David Eagleman compares it to the CEO of a big company who believes he is making all the decisions in the company (Eagleman, 2011). The reality is that a large company requires a huge number of decisions be carried out at a lower level than the CEO, because they cannot handle everything. For tough decisions, the CEO's decisions are required to make a call. He/she receives technical reports and advice from experts in the company recommending consideration of the value of different options. As the CEO decides, most of the data analysis is already completed before he has even looked at the reports.

This metaphor is quite handy. The CEO's time is restricted and thus should only be used to solve complex/new situations; while lower divisions should solve easy/ usual problems. In the same way, our ability to process information consciously is much more limited than our ability to do it unconsciously. The reason seems to be that while separate areas of the brain can process specific information, the phenomenon of consciousness arises from activation of the whole brain (Dehaene, 2014). Consciousness seems to arise as the brain integrates information from its different parts, sends it to other areas to compare and cross validate different types of information. As this activation takes place over the whole brain, it can only solve one given problem at a time.

Going back to the CEO metaphor, once his/her attention has been raised to a problem, the CEO will integrate the different inputs from different technical reports, compare them, and possibly ask an advisor in another area to interpret the information received from a previous advisor. This process is slow, but necessary to avoid mistakes in new situations. If the problem at hand becomes a regular one, new rules of decision-making can be established for the decision to be sent to a lower division.

From this visualization, it is important to recognize that most of the legwork is completed outside of consciousness. In most cases, we do not choose an option; we have rationally assessed its merit through a subconcious quantitative assessment. The estimation of the option value has been done behind the realm of our conscious-ness by brain processes that have cross-referenced the characteristics of the options with characteristics of past experiences and hardwired patterns. If an option's value is clearly better than others are, we can choose it without barely thinking about it. If several options have competing values, we may have mixed feelings, which will lead us to ponder consciously about the problem to compare the relative weight of evidence in favour of each option. In this decision-making process, our decision is not purely driven by objective reasons which are consciously voiced in our head, instead, it is driven by a stream of feelings which represent information which has already been processed by our brain beyond our consciousness.

Psychologists and neuroscientists have shown that, in many cases, the conscious reasons are made up after the decision is made in the brain to make sense of it *ex post* (Mercier & Sperber, 2011). For instance, through brain stimulation, a participant can be prompted to make a given choice in an experiment. When questioned why he made this choice, the participant will typically give reasons as though these had motivated his choice (Ramachandran, 2012).

These elements show that we are inclined to overstate the importance of the voice in our head as the center of our decision-making process. Our feelings, intu-itions, and emotions are not irrelevant or even irrational elements; they are valuable inputs in our decision-making process and they are the outcome of computations processed subconsciously.

Not only do they provide the required preprocessed information for our con-scious decision-making to take place, they do much more. At any moment in time, multiple brain processes treat information and assess whether it should be sent to consciousness to be treated. Even though we do not think consciously of things, our brain keeps scanning the available information. While consciousness only focuses on one question at a time, our brain is considering much more information to assess whether other issues must be considered.

If the weight of evidence is compelling enough, a new issue can supersede a previous one in our consciousness in the same way that an advisor can convince the CEO to look at new information.

Intuition in Financial Investment

The model of human cognition described above gives a new vision of the role of intuition and analytics. Intuitions are not irrational flukes that need to be kept in check. Traders should not primarily mistrust their feelings. Instead, they need to understand them to be able to make use of them.

If a trip to the supermarket is a complex task, investing on the financial market is another challenge altogether. Financial decisions involve making judgments about forecasting uncertainty. Traders must form a view on the likely evolution of a wide range of economic, political and social phenomena that may affect the future value of assets. Furthermore, financial markets are characterized by even more complex uncertainty: the strategic uncertainty, which arises from the actions of other financial agents. A trader needs to anticipate the psychology of other traders and form best response strategies.

The complexity of financial decisions means that there are clear benefits from analytic tools. Modern computers can process enormous amount of data in seconds, solve optimisation problems and use quantitative models that contain accumulated insights from the past. Analytics is therefore a critical helping tool to survive and thrive in modern financial markets. They should not however lead to reject the role of intuitions.

Our evolved ape brain was not designed to solve financial problems. It is therefore not surprising that behavioral finance has listed a wide range of suboptimal behavior on financial markets. The evidence of the existence of irrational exuberance in times of booms and bust comes in direct contradiction with the view of cold rational traders (Shiller, 2005; Newell & Page, 2017) and progress in neuroscience has shown roots of such behavior (Coates & Herbert, 2008; Coates & Page, 2016; Kandasamy et al., 2014).

However, it would be an error to put away all our intuitions and emotions. These are key parts of our decision process. They must play a role because analytics cannot, in most cases, provide a definite answer. If finding optimal financial decisions is beyond a human brain, the truth is that it is beyond the ability of computers as well. The complexity of the real world is hard to understate. Analytic techniques can only provide a helping hand by relying on great simplifications of the world through models. Consequently, there is a gap between the real world and the models used by analytic methods. In many situations, this gap may be small and analytic methods provide good answers quickly. In other situations, the match between models and reality may become more uncertain and therefore relying purely on analytics may not be as fruitful or even dangerous.

The human brain has a key advantage relative to computers; its ability to detect and makes sense of complex patterns. It is this ability that allowed a player like Kasparov to compete with Big Blue. Even though the program could estimate 200 million positions per second versus 3 for Kasparov, his human brain could associate the patterns on the board with strategic perspectives and values that are very hard for a computer to assess (Silver, 2015).

Similarly, on a financial market, an investor can appreciate the complex patterns and make sense of them in ways a computer cannot. Using experience, he/she can get a feeling of what is likely to happen next and how to adjust decisions suggested by analytic techniques.

Gut feelings do not come naturally. The gut feelings of a naïve investor are likely to be of little help. However, our brain is also not designed to play chess. Through learning and experience, the expert chess player can acquire the intuition/gut feelings, which will efficiently guide his decisions. Similarly, a financial investor, through experience and learning can be able to have intuitions that valuably complement analytic prediction. Indeed, we have found that traders Sharpe ratio tends to increase over time with experience (Coates & Page, 2009).

The positive intuition of an experienced trader will reflect a detection of patterns whose characteristics have been cross-referenced with experience, leading his/her brain to recognize more positive than negative opportunities. In a world where it is not good enough to be good, the best performing traders will be those who can harness the insights of these intuitions to make sense of and benefit from them (Coates, 2012).

Case Study: Interoceptive Ability and Trader's Performance*

It is with this perspective that we designed a study to investigate the role of gut feelings in financial success (Kandasamy et al., 2016). We hypothesized that the ability of traders to understand and use their gut feelings can be linked with their ability to ready their physiological signals, which are associated with our emotions. Studies had found that measures of interoceptive ability (the sense of our internal body activity) were associated with superior performance to some laboratory gambling tasks (Dunn et al., 2010; Sokol-Hessner, Hartley, Hamilton, & Phelps, 2015). However, little research has been done on real-world decision-making.

---

* This section largely reprints content from the study "Interoceptive Ability Predicts Survival on a London Trading Floor" (Kandasamy et al., 2016).

We conducted a study on the trading floor in a London hedge fund to test the hypothesis that traders with greater ability to sense signals from their bodies would make more money.

Interoceptive Ability

Research has shown that people vary in their ability to generate and sense interoceptive signals. Physiological tests have been developed to measure this ability. The most common are heartbeat detection tests (Herbert, Pollatos, & Schandry, 2007; Katkin, Wiens, & Öhman, 2001).

In this study, we used two heartbeat detection tests. First, each trader performed a heartbeat counting task, in which he was instructed to silently count, without touching chest or any pulse point, how many heartbeats he felt over brief time intervals (Schandry, 1981). This counting task was repeated six times, using randomized time periods of 25, 30, 35, 40, 45, and 50 seconds. The number of heartbeats perceived by the trader within each period was compared with the actual number of heartbeats, as recorded over that interval objectively and noninvasively by pulse oximetry. A heartbeat detection accuracy score was expressed as a percentage using the following equation (Hart, McGowan, Minati, & Critchley, 2013):

$$\text{Heartbeat detection score} = \left(\left[1 - \left[\text{absolute value of } (\text{actual} - \text{estimated}) \div \right. \right. \right.$$
$$\left. \left. \left. \text{mean of } (\text{actual} + \text{estimated}) \right] \times 100 \right] \right)$$

After each trial of the counting task, the traders rated their confidence in their heartbeat detection using a visual analogue scale. At one end of this scale was marked "Total guess/No heartbeat awareness," at the other end "Complete confidence/Full perception of heartbeat."

In the second heartbeat detection task, traders were asked to judge whether a series of ten auditory tones was played synchronously with, or delayed relative to, their own heartbeats (Brener, Liu, & Ring, 1993; Whitehead, Drescher, Heiman, & Blackwell, 1977). The tones (presented at 440 Hz and lasting 100 ms) were triggered by individual heartbeats. On synchronous trials, the 10 notes occurred at the rising edge of the finger pulse pressure wave; on delay trials, they followed 300 ms later. At the end of each trial, the participant indicated whether the tones were synchronized or not. Each trader performed 15 trials of this task. Their score was the percentage of right answers. This task was performed after the counting task to prevent the traders from developing any explicit knowledge about their own heart-rates (Ring, Brener, Knapp, & Mailloux, 2015).

Data was also collected on each trader's age, years of trading, and profit and loss statement, known as their P&L. Comparative data from a control group of 48 high-functioning (mostly students and postgraduates) nontrader males, matched with traders on age, were drawn from participants who had performed identical

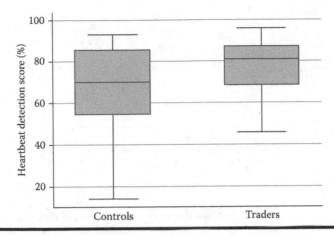

Figure 2.1   Box plots showing that mean interoceptive accuracy (score on heart-beat counting task) for traders (N = 18) was significantly higher than for a cohort of nontraders (N = 48).

interoceptive tests (same equipment and procedures) in studies undertaken at the University of Sussex (Garfinkel, Seth, Barrett, Suzuki, & Critchley, 2015).

Traders Have Enhanced Interoceptive Ability Compared to Nontrading Individuals

As a first step in assessing the importance of interoception in financial risk taking, we tested simply whether traders scored higher on heartbeat detection tasks than nontrading controls. We compared the average score on the heartbeat counting task for the traders with the average score from our control group. We found that the traders had significantly higher scores, indicating greater interoceptive accuracy, than the controls, with a mean score of 78.2 for traders and 66.9 for controls ($p = 0.011$, N = 66, Figure 2.1).

Interoceptive Ability Predicts Trader Profitability

To test the hypothesis that interoceptive accuracy predicts trading performance, we collected data on each trader's average daily profit and loss (P&L) over the previous year. One of the traders was acting as trading manager, so he did not have his own P&L; another trader had a P&L more than 5 standard deviations higher than the mean of other traders, so even with data transformations he distorted the distribution. Both traders were therefore omitted from analyses that used "raw" P&L. We found that the traders scores on the heartbeat counting task predicted their P&L (coeff = 2.61, $R^2 = 0.27$, $p = 0.007$, N = 16). To further validate this association, we also rank-ordered the P&L, which permitted the inclusion of the trader with

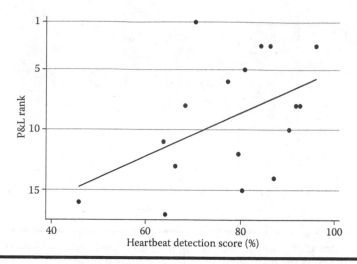

Figure 2.2   Regression line plotting score on the heartbeat counting task against the traders rank-ordered P&L, with 1 representing the most profitable trader, 17 the least.

the outlying P&L. Again, we found that heartbeat detection score predicted relative trading performance (coeff = 17.84, R^2 = 0.22, p = 0.01, N = 17, Figure 2.2).

Interoceptive Ability Predicts Survival in the Financial Markets

If heartbeat detection score predicts traders P&L, does it also predict how long traders survive in the financial markets? To answer this question, we plotted heartbeat detection scores against years of experience in the financial markets and found that a trader's heartbeat counting score predicted the number of years he had survived as a trader (coeff = 21.64, R^2 = 0.344, p = 0.001, N = 18, Figure 2.3).

The solid red line is the regression plot. A regression model with conditional mean and conditional standard deviation (std) estimated jointly is used to assess the significance of changes in the heartbeat detection mean and STD over years of trading. Light dashed horizontal lines are +/−1 std. Vertical dotted lines show distributions of residuals for each bucket of trading experience. These distributions show a declining variance of heartbeat detection as years of experience increase.

Does this result mean the markets select for traders with greater interoceptive ability? To pursue this analysis, we reasoned as follows: (1) If firms know nothing about interoception (and the hosting firm of our study did not) they will not hire traders based on heartbeat detection scores. We should find, therefore, that beginner traders and the nontrading population have mean heartbeat detection scores and a standard deviation of scores that do not differ significantly. (2) If the market selects for traders with good gut feelings (i.e., high heartbeat detection scores), then

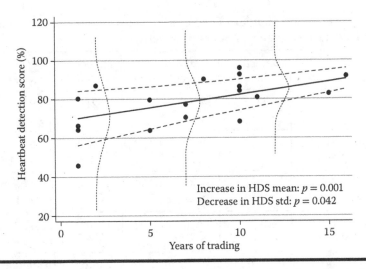

Figure 2.3 Years of trading experience plotted against heartbeat detection score (HDS).

what we should find is that as the traders' careers progress and the market selects for traders with better gut feelings, those with low heartbeat detection scores will be eliminated from the markets. Average heartbeat detection scores for the traders will rise, while the standard deviation of scores will fall as the lower end of the range drops out.

. To test this possibility, we partitioned the group according to experience: junior traders (1–4 yrs experience), mid (5–8 yrs), and senior (>8 yrs), and then calculated the mean and standard deviation of detection scores for each bucket (Table 2.1). We found that beginner traders did not differ significantly from controls in the mean ($p = 0.852$, N = 53) nor standard deviation ($p = 0.614$, N = 53) of their heartbeat detection score. Over time, however, the traders mean heartbeat detection score did indeed rise, from 68.7 for beginners to 85.3 for experienced traders. This latter level differed significantly from controls (T test, $p = 0.02$, N = 56). In addition, the standard deviation decreased, from 16 for beginners to 8.6 for experienced traders, and this latter standard deviation also differed significantly from controls (F test, $p = 0.018$, N = 56) (Table 2.1). This result is represented graphically in Figure 2.3, where the variance of residuals declines with the number of years of trading (coeff = -0.063, $p = 0.042$, N = 18), the pattern of these residuals forming a wedge shape rather than the customary parallelogram.

Interoceptive Accuracy But Not Confidence Informs Trading Performance

Next, we analyzed the traders level of confidence in their estimates of their own heartbeats, as recorded on visual analogue scales during the heartbeat counting task

Table 2.1 Mean and Standard Deviation of Detection Scores for Controls and for Traders Organized by Years of Experience

Cohort	Detection Mean	Difference from Controls	Detection Std	Difference from Controls
CONTROLS (N = 48)	66.9		21.3	–
JUNIOR TRADERS 1–4 YRS (N = 5)	68.7	+1.8 ($p = 0.852$, N = 53)	16.0	–5.3 ($p = 0.614$, N = 53)
MID TRADERS 5–8 YRS (N = 5)	76.3	+9.4 ($p = 0.339$, N = 53)	9.9	–11.4 ($p = 0.144$, N = 53)
SENIOR TRADERS >8 YRS (N = 8)	85.3	+18.4 ($p = 0.02$, N = 56)	8.6	–12.7 ($p = 0.018$, N = 56)

Note: T tests are used to compare means, and F tests of equality of standard deviation are used to compare standard deviations.

(Ceunen, Van Diest, & Vlaeyen, 2013; Garfinkel et al., 2015). We found no significant correlations between confidence and heartbeat detection accuracy (coeff = 0.17, $p = 0.51$, N = 17); nor between confidence and P&L (coeff = –0.01, $p = 0.97$, N = 17); nor between confidence and years of survival (coeff = –0.16, $p = 0.54$, N = 17).

This disjunct between the traders heartbeat detection accuracy and their confidence in their accuracy may seem contradictory but, as with many physiological measures, objective performance and subjective appraisal often diverge. In the case of interoceptive ability, it is conceivable that a self-conscious awareness of interoceptive signals impairs the signal's utility because people may dismiss the signals as "merely" physical and distracting. Alternatively, self-consciousness may impair risk taking in much the same way that focusing self-consciously on, say, your tennis stroke can impair your game. Related to this point, it is noteworthy that the control group in our study included 16 medical students who, despite their more advanced understanding of the cardiovascular system, performed on average worse than the traders on the counting task (65.9 versus 78.2), although this difference only approached significance ($p = 0.056$, N = 34).

Conclusion

In the old dichotomies, mind/body and rationality/passions, a coldly rational decision maker can abstract himself/herself from the influences of emotions and visceral feelings to engineer good decisions. Behavioral economics has shown that this vision of a rational decision maker did not stand as a positive theory of decision; that is, it does not correspond to the reality of how people make decisions.

However, behavioral economists from the "heuristics and biases" school like Kahneman and Tversky implicitly kept the model of homo economicus as a normative model of what is a good decision. Human decision-making processes were in comparison considered as imperfect and flawed.

New research connecting behavioral economics and neuroscience now gives another picture whereby humans are better decision makers than we thought; but many of the decisions derive from brain processes beyond the realm of consciousness. Faced with a highly complex environment, the brain optimally allocates its scarce computational resources and only thin slices of reality are acquiesced to a conscious analysis. Real-time decisions require the brain to rely on a wide range of automated processes to treat information and provide quick evaluations relevant for the value of different options considered. Given this large activity, a lot of what we perceive from reality gives rise to feelings which often only hover at the limit of consciousness. These feelings that present as intuition, and gut feelings are not irrational or irrelevant. They contain real information.

In the world of finance where agents are in a race to make the best decisions, dismissing these intuitions would be dismissing a lot of potentially valuable input in the decision-making process. Top traders certainly achieve high performance by fine-tuning their intuitions through extensive experience. While our gut feeling mechanisms have not been designed to help us operate in financial markets, accumulated experience can be encoded in gut feelings, which can be helpful. Having informative gut feelings is not enough though. For them to play a positive role in financial decision-making, top traders must be able to recognize them and channel this information in decisions.

To test such an idea, I will report the result from a study conducted with colleagues, which found that the ability to sense internal body signals, interoception, is positively associated with performance on the trading floor. We conjecture that by being able to recognize the somatic markers associated with emotions, traders can make better decisions in a highly complex environment.

This research suggests that intuition, or gut feeling, should be valued on the trading floor as a complement of analytic tools. Like any signal, intuitions can be mistaken, but experienced traders will have encoded a wide range of experience and memories that will help them make sense of new situations, feeding them intuitions about the value of different choices. When making decisions, traders should be ready and willing to use both types of input: analytics and intuitions. However, good traders will know that no tool should supersede the other one fully. Rather, intuitions should help question analytics, and vice versa. In many situations where analytic models may be lacking realism, intuition can fruitfully contribute to appreciate the degree of confidence to give an analytic prescription; and whether one should deviate from them. In situations where analytics conflict with intuitions, one should consider questioning these intuitions to make sure they do not take precedence over good data. Such a way to proceed, is like aiming to look at a chess game like Kasparov, while being able to run the millions of Deep Blue simulations as a helping tool. The computer simulations are a great help, but, for the time being, there are things that the human brain does which are beyond the scope of computer programs.

References

Bechara, A., Damasio, A. R., Damasio, H., & Anderson, S. W. (1994). Insensitivity to future consequences following damage to human prefrontal cortex. *Cognition, 50*(1–3), 7–15. doi: 10.1016/0010-0277(94)90018-3.

Brener, J., Liu, X., & Ring, C. (1993). A method of constant stimuli for examining heart-beat detection: Comparison with the Brener-Kluvitse and Whitehead methods. *Psychophysiology, 30*(6), 657–665. doi:10.1111/j.1469-8986.1993.tb02091.x.

Bruni, L., & Sugden, R. (2007). The road not taken: How psychology was removed from economics, and how it might be brought back. *The Economic Journal, 117*(516), 146–173. doi:10.1111/j.1468-0297.2007.02005.x.

Ceunen, E., Van Diest, I., & Vlaeyen, J. W. S. (2013). Accuracy and awareness of perception: Related, yet distinct (commentary on Herbert et al., 2012). *Biological Psychology, 92*(2), 426–427. doi:10.1016/j.biopsycho.2012.09.012.

Coates, J. M. (2012). *Hour between dog and wolf.* New York, NY: Penguin Group. Retrieved from http://www.myilibrary.com?id=710147.

Coates, J. M., & Herbert, J. (2008). Endogenous steroids and financial risk taking on a London trading floor. *Proceedings of the National Academy of Sciences, 105*(16), 6167–6172. doi:10.1073/pnas.0704025105.

Coates, J. M., & Page, L. (2009). A note on trader Sharpe Ratios. *PLOS ONE, 4*(11), e8036. doi:10.1371/journal.pone.0008036.

Coates, J. M., & Page, L. (2016). Biology of financial market instability. In S. N. Durlauf & L. E. Blume (Eds.), *The new Palgrave dictionary of economics.* London: Palgrave Macmillan.

Damasio, A. R. (2006). *Descartes' error: Emotion, reason and the human brain.* London: Vintage Books. Retrieved from https://www.overdrive.com/search?q=E0BCCECB-B8CC-4CCE-A5D0-7A3EE6CFFE74.

Dehaene, S. (2014). *Consciousness and the brain: Deciphering how the brain codes our thoughts.* New York: Penguin.

Dunn, B. D., Galton, H. C., Morgan, R., Evans, D., Oliver, C., Meyer, M., . . . Dalgleish, T. (2010). Listening to your heart: How interoception shapes emotion experience and intuitive decision making. *Psychological Science, 21*(12), 1835–1844. doi:10.1177/0956797610389191.

Eagleman, D. (2011). *Incognito: The secret lives of the brain.* Melbourne: The Text Publishing Company. Retrieved from https://www.overdrive.com/search?q=ADF7CF30-BF23-4D36-9508-1934B00385A4.

Freud, S., & Strachey, J. (1977). *Introductory lectures on psychoanalysis.* New York, NY: Norton.

Garfinkel, S. N., Seth, A. K., Barrett, A. B., Suzuki, K., & Critchley, H. D. (2015). Knowing your own heart: Distinguishing interoceptive accuracy from interoceptive awareness. *Biological Psychology, 104*, 65–74. doi:10.1016/j.biopsycho.2014.11.004.

Gigerenzer, G. (2000). *Simple heuristics that make us smart.* New York, NY: Oxford University Press. Retrieved from http://public.eblib.com/choice/publicfullrecord.aspx?p=279529.

Gintis, H. (2007). A framework for the unification of the behavioral sciences. *Behavioral and Brain Sciences, 30*(01). doi:10.1017/S0140525X07000581.

Gladwell, M. (2007). *Blink: The power of thinking without thinking* (1st Back Bay trade pbk. ed.). New York, NY: Back Bay Books.

Hahn, H., Neurath, O., Carnap, R. (1996). The scientific conception of the world. The Vienna circle. In *The emergence of logical empiricism: From 1900 to the Vienna circle* (pp. 321–340). New York, NY: Garland Publishing.

Hart, N., McGowan, J., Minati, L., & Critchley, H. D. (2013). Emotional regulation and bodily sensation: Interoceptive awareness is intact in borderline personality disorder. *Journal of Personality Disorders, 27*(4), 506–518. doi:10.1521/pedi_2012_26_049.

Herbert, B. M., Pollatos, O., & Schandry, R. (2007). Interoceptive sensitivity and emotion processing: An EEG study. *International Journal of Psychophysiology, 65*(3), 214–227. doi:10.1016/j.ijpsycho.2007.04.007.

Kahneman, D. (2012). *Thinking, fast and slow.* London: Penguin.

Kahneman, D. (2003). Maps of bounded rationality: Psychology for behavioral economics. *The American Economic Review,* 93(5): 1449-1475.

Kandasamy, N., Garfinkel, S. N., Page, L., Hardy, B., Critchley, H. D., Gurnell, M., & Coates, J. M. (2016). Interoceptive ability predicts survival on a London trading floor. *Scientific Reports, 6*(1), 1–6. doi:10.1038/srep32986.

Kandasamy, N., Hardy, B., Page, L., Schaffner, M., Graggaber, J., Powlson, A. S., . . . Coates, J. (2014). Cortisol shifts financial risk preferences. *Proceedings of the National Academy of Sciences, 111*(9), 3608–3613. doi:10.1073/pnas.1317908111.

Katkin, E. S., Wiens, S., & Öhman, A. (2001). Nonconscious fear conditioning, visceral perception, and the development of gut feelings. *Psychological Science, 12*(5), 366–370. doi:10.1111/1467-9280.00368.

Mercier, H., & Sperber, D. (2011). Why do humans reason? Arguments for an argumentative theory. *Behavioral and Brain Sciences, 34*(02), 57–74. doi:10.1017/S0140525X10000968.

Montague, R. (2007). *Your brain is almost perfect: How we make decisions.* New York, NY: Plume.

Newell, A., & Page, L. (2017). Countercyclical risk aversion and self-reinforcing feedback loops in experimental asset markets. No. 050. *QUT Business School.*

Ramachandran, V. S. (2012). *The tell-tale brain: A neuroscientist's quest for what makes us human.* New York, NY: W. W. Norton.

Ring, C., Brener, J., Knapp, K., & Mailloux, J. (2015). Effects of heartbeat feedback on beliefs about heart rate and heartbeat counting: A cautionary tale about interoceptive awareness. *Biological Psychology, 104*, 193–198. doi:10.1016/j.biopsycho.2014.12.010.

Schandry, R. (1981). Heart beat perception and emotional experience. *Psychophysiology, 18*(4), 483–488. doi:10.1111/j.1469-8986.1981.tb02486.x.

Shiller, R. J. (2005). *Irrational exuberance.* Princeton University Press.

Silver, N. (2015). *The signal and the noise: Why so many predictions fail - but some don't.* New York, NY: Penguin Books.

Sokol-Hessner, P., Hartley, C. A., Hamilton, J. R., & Phelps, E. A. (2015). Interoceptive ability predicts aversion to losses. *Cognition and Emotion, 29*(4), 695–701. doi:10.1080/02699931.2014.925426.

Stigler, G. J., & Becker, G. S. (1977). De gustibus non est disputandum. *The American Economic Review, 67*(2), 76–90..

Whitehead, W. E., Drescher, V. M., Heiman, P., & Blackwell, B. (1977). Relation of heart rate control to heartbeat perception. *Biofeedback and Self-Regulation, 2*(4), 371–392. doi:10.1007/BF00998623.

Chapter 3

Data and Analytics: A Matter of Trust

Wilds Ross

Contents

Data analysis to support business decisions is nothing new; however, the application of new technologies enabled by the constant development in computing power, and the discovery or re-discovery of innovative techniques made practical by these developments is taking the business community by storm. With technological acceleration, data and analytics (D&A) activities are subject to the adoption cycles of both technology and methodology. Technological adoption often follows the well-established normal growth curve first observed in the 1950s and includes innovators, early adopters, the majority adopters, and laggards. While methodology adoption follows a similar pattern with process-unique drivers for acceptance, in the business community, both adoption patterns have a common basis before becoming ubiquitous: trust.

This is to be expected when anything challenges the status quo. Data and analytics is replacing biased, gut-feel, and subjective decision-making with objective, data-driven prescriptive insights that allow organizations to better serve customers, drive operational efficiencies, and secure their organization from both internal and external risks. Acting on findings from D&A activity requires a belief that the results represent reality, and accurately predict the behavior of the business. This is not easily done when the results are obscured within statistics and the abstract language of data science, or generated by teams of people that have limited experience within the business domain and thus may not have the expected level of alignment to business activity that is necessary for full realization. Big Data, data science, machine learning, and other analytics topics are not necessarily common currency for many executives that are accustomed to developing their own confidence by examining traditional analysis methods with minimal effort. Coupled with a legacy of insecurity in the data, this has led to a trust gap, whereby some businesses have been reluctant to adopt the findings and benefit from D&A results. This is far from universal, however. The businesses that have developed effective D&A programs and can trust in what they discover have achieved success, and in many cases, superior data and analytics and the confidence to act upon it have been credited with significant effects toward business performance.

D&A are often deployed in support of three primary areas of business focus: growth, operational management, and compliance.

Helping to better understand customers and create new experiences

In today's business ecosystem, customers are in the driver's seat. Empowered by digital information and resources, customers will defect if a brand doesn't meet their every need. To remain competitive in this environment, organizations must find new ways to deliver value to customers and prospects. So, it comes as no surprise that businesses rely on D&A capabilities to create new sources of value for customers. Among other things, D&A is integral to understanding how products are being used, understanding existing customers better, and developing new products and services.

Streamlining existing operations

As firms become more efficient in existing processes, they open opportunities to invest time and budget in things that will make them more competitive. D&A is essential to gaining better visibility into business performance, driving process and cost efficiency, and driving strategy and change.

Managing risk and compliance

While there is a lot of upside potential to firms advancing their use of D&A, managing the downside risks is imperative. For example, D&A is critical for spotting fraud, and identifying and managing other business risks. For companies that need to comply with industry regulations, D&A helps them meet their reporting and transparency requirements. Beyond compliance, an important consideration for how firms use D&A is customer trust.

These areas are common to most businesses, and D&A is becoming more prevalent across a wide range of industries, but as its importance grows, businesses are faced with significant challenges, and these issues revolve around trust.

A Matter of Trust

In a global environment defined by constant disruption, business leaders need to be confident in their decisions. This means being able to trust their data, their algorithms and their analytics capabilities. However, significant questions are now emerging about the trust placed in the data, the analytics and the controls that underwrite a new way of making decisions.

How do we know a result is right or that automated decisions are doing the right thing? What does "right" mean and to what extent does it matter? Who should be the judge? As analytics goes mainstream, questions surrounding trust will continue to evolve; not only across most sectors, but also for regulators, policymakers and those who safeguard consumer rights.

Executives cannot make sound business decisions if they don't trust their data and analytics. Yet according to research by KPMG,* only around a third of organizations have a high level of confidence in their customer insights or the analytics they receive on their business operations.

As analytics increasingly drive the decisions that affect us as individuals, as businesses and as societies, there must be a heightened focus on ensuring the highest level of trust in the data, the analytics and the controls that generate desired outcomes.

To deliver business results, organizations must think about D&A trust not as a nice-to-have, but as a strategic way to bridge the gap among decision makers, data scientists, and customers.

In reality, trust is not at the forefront of many leaders' minds. While some mistrust the input or the output, others often don't care about trust until they are forced to. Unfortunately, by then, the damage could have already been done through poor and costly decisions or, even worse, through hurting the organization's reputation and brand with its customers.

At KPMG, we have spent some time analyzing the current trust gap affecting organizations around the world and developed a framework for assessing and building trust in D&A.

Trust in D&A should be nonnegotiable. Yet, our research of over 2,000 organizations revealed a trust gap. Organizations not only lack confidence in their ability to effectively use and link D&A to positive business outcomes, they also highlighted a lack of trustworthiness in the analytics used to drive decision-making.

---

* "Building Trust in Analytics: Breaking the Cycle of Mistrust in D&A," KPMG LLP, October 2016. https://home.kpmg.com/xx/en/home/insights/2016/10/building-trust-in-analytics.html

What is the result? Few firms understand whether the D&A models that are being created achieve what they were intended to. This creates a cycle of mistrust, which has an impact on future analytical investments and their return on investment (ROI).

The potential impact of mistrust in D&A can be seen beyond the four walls of an organization. As customers increasingly rely on D&A to make contextually relevant and difficult decisions, any mistrust from the organization can affect a brand's reputation, and ultimately, its bottom line.

The trust gap has implications beyond the walls of an organization. Today, complex analytics underpin many important decisions that affect us as individuals, as businesses, and as societies. With so much riding on the output of D&A, significant questions are starting to emerge about the trust that we place in the data, analytics, and the controls that underwrite a new way of making decisions.

When consumers are involved, the absence—or loss—of trust can clearly have a major impact on the success of a brand's reputation and sales performance. Anecdotal evidence over the last few years suggests that data breaches of some of the world's leading retailers which exposed millions of shoppers' payment cards had a major impact on brand reputation and in some cases, financial results. Lack of trust and misplaced trust can also have a wider impact on society. For example, the financial crisis of 2008 was exacerbated by predictive risk models which were technically "correct," but failed spectacularly in their intended purpose. Making decisions or targeting consumers based on inaccurate predictions will quickly erode, if not extinguish, consumer trust and shake the confidence of those executives who rely on these predictions to make informed decisions.

At the core of trusted analytics are rigorous strategies and processes that aim to maximize trust. Some are well known but challenging, such as data quality. Others are relatively new and undefined in the D&A sphere, such as ethics and integrity. KPMG believes that organizations should take a systematic approach to trust that spans the life cycle of analytics and has outlined a framework that consists of four *"anchors."*

"Anchors" of Trust

Organizations must strive to engender trust and accelerate the time-to-value for all D&A projects. To close the gap, KPMG has outlined a framework of four anchors of trust that underpin trusted analytics: (1) quality, (2) effectiveness, (3) integrity, and (4) resilience. Using these anchors, we outline the processes, practices, and governance that all organizations must implement to deliver trusted data, analytics, and models among decision makers, data scientists, and customers.

Let's look at what each of the anchors means.

> **Quality:** Are the fundamental building blocks of D&A good enough? How well do organizations understand the role of quality in their current approach to organizing, developing, and managing data and analytics?

Effectiveness: Do the analytics work as intended? Can organizations determine the accuracy and utility of the outputs?

Integrity: Is the use of D&A considered acceptable? How well aligned is the organization with analytics regulatory compliance and privacy/ethical needs? Is D&A transparently used both internally and externally?

Resilience: Are long-term operations optimized? How good is the organization at ensuring governance and security throughout the analytics life cycle?

Each anchor of trust is relevant at every stage of the D&A life cycle—from data sourcing, to data preparation and blending, to analysis and modeling, to usage and deployment, and finally through to measuring effectiveness and back to the beginning of the cycle.

Breaking the Cycle of Mistrust

Firms need to break the cycle of mistrust, and to do that they need to close several gaps in D&A capabilities that underpin the anchors of trust. Our study looked at core capabilities within each of the anchors and asked D&A decision makers to rate how well their organizations align to each of those capabilities. Unfortunately, many firms are ill-equipped to deliver trusted analytics due to significant gaps in capabilities. In fact, apart from D&A regulatory compliance, where organizations tended to perform strongest, they struggled to achieve excellence across each of the D&A anchors. The study showed that only 10% of organizations believed that they excelled in developing and managing D&A. Additionally, a mere 13% excelled in the privacy and ethical use of D&A, and less than a fifth (16%) performed well in ensuring the accuracy of models they produce. The following sections explore each of the anchors in more detail.

Quality

Quality is the trust anchor most commonly cited by internal decision makers. Most organizations understand and struggle with data quality standards for accuracy, completeness and timeliness. As data volumes increase, new uses emerge and regulation grows, the challenge will only increase. Everybody recognizes that at some levels, all analytical models are "wrong" and not a perfect reflection of reality. But where does it matter most?

To drive quality in D&A, organizations need to ensure that both the inputs and development processes for D&A meet the quality standards that are appropriate for the context in which the analytics will be used. In many organizations, questions are raised about choice of data sources and data "lineage" (i.e., where the data originated and what process it took to arrive as input data to a system or decision engine).

The quality of analytics poses huge potential trust issues. Statistical and algorithm design, model development approaches and quality assurance are becoming critical. Organizations are struggling to assess quality in scenarios in which the impact of low quality can be high or where there is no known right answer with which to compare the output of a new decision engine.

Effectiveness

When it comes to D&A, effectiveness is all about real-world performance. It means that the outputs of models work as intended and deliver value to the organization. This is the top concern of those who invest in D&A solutions, both internal and external to the organizations.

The problem is that D&A effectiveness is becoming increasingly difficult to measure. In part, this is because D&A is becoming more complex and therefore the "distance" between the upstream investment in people and raw data is often far removed from the downstream value to the organization.

When organizations are not able to assess and measure the effectiveness of their D&A, it becomes easy for those making decisions to miss the full value of their investments and assume that a large proportion of their D&A projects "don't work." This, in turn, erodes trust and limits long term investment and innovation.

Organizations that can assess and validate the effectiveness of their analytics in supporting decision-making can have a huge impact on trust at the board level. The corollary of this, of course, is that organizations that invest without understanding the effectiveness of D&A may not move the needle on trust or value at all.

Integrity

Integrity can be a difficult concept to pin down. In the context of trusted analytics, we use the term to refer to the acceptable use of D&A, from compliance with regulations and laws such as data privacy through to less clear issues surrounding the ethical use of D&A such as profiling. This anchor is typically the top concern of consumers and the public, generally.

Behind this definition is the principle that with power comes responsibility. Algorithms are becoming more powerful, and can have hidden or unintended consequences. How do we decide what is acceptable and what isn't? Where exactly does accountability lie, and how far does it extend?

This is a new, uncertain and rapidly changing anchor of trust with few globally agreed best practices. Individual views vary widely, and there is often no right answer. Yet integrity has a high media profile and has potentially enormous implications, not only for internal trust in D&A, but also for public trust in the reputation of any organization that gets it wrong.

Integrity goes beyond consumer trust issues. Most organizations understand that D&A offers huge potential benefits by replicating good decisions and limiting human inconsistencies and biases. If algorithms are well "trained," then race or gender biases, for example, can be removed. It stands to reason, therefore, that an effective combination of human and machine can offer fairer, more trusted decisions.

However, this is not guaranteed. If not well managed throughout the D&A life cycle, algorithms can also introduce unintentional, hidden biases because of the data on which they have been trained. Automated decision engines can also make the ethical consequences feel emotionally distant to the humans who are nominally accountable.

Resilience

Resilience in this context is about optimization for the long term in the face of challenges and changes. Cyber security is the best-known issue here, but resilience is broader than information security. Failure of this trust anchor undermines all the previous three.

Unlike traditional software, applications which apply machine learning and operate in a complex D&A ecosystem with fast-changing data sources are likely to change their function, impact, and value throughout their operational lifetime, sometimes quite suddenly.

Basic resilience is key to winning customer trust. It only takes one service outage or one data leak for consumers to quickly move to (what they perceive to be) a more secure competitor. It also only takes one Big Data leak for the regulators to come knocking and for fines to start flying. Strong governance and control can also help reduce duplication of effort and therefore, help improve the value of D&A across the enterprise.

Release of new data sources by third parties can also have unintended impacts on existing analytics and the art of the possible. In 2009, for example, academics In the United States demonstrated that it was possible to predict an individual's social security number with remarkable accuracy using only public data such as the Social Security Administration's Death Master File and other personal data such as profiles on social networking sites.*

Recommendations for Building Trusted Analytics

Trust is not a project. Strengthening the anchors of trust is not a one-time exercise or a compliance checklist. It is a continuous endeavor that should span the entire enterprise. From the sourcing and preparation of data through to the outcomes and measurement of value, building trust in analytics requires executives to look across their D&A life cycle, from data through to insights and ultimately to generating value. There are no roadmaps for driving trust, no software solutions or perfect

---

* http://www.pnas.org/content/106/27/10975.ful

answers. However, there are some best practices that all organizations can consider and adopt.

Here are seven ideas that should help any organization create their own approach to building D&A trust.

1. **Start with the basics—Assess your trust gaps:** Undertake an initial assessment to see where trusted analytics is most critical to your business, and then focus on those areas. Key risks can often be reduced with some very straightforward changes, such as the use of simple checklists.

2. **Create purpose:** Clarify and align goals. Ensure that the purpose for your data collection and the associated analytics is clearly stated. Make D&A performance and impact measurable. The aims and incentives of the D&A "owners" should align with the goals of its users and with those who could be affected by it. Lack of clarity around purpose and misalignment of D&A goals can create mistrust, dilute ROI and open the door to inadvertent misuse.

3. **Raise awareness:** Increase internal engagement. Building awareness and understanding of D&A among business users is critical to breaking the cycle of mistrust. Involve key stakeholders and establish multidisciplinary project teams, combining D&A leaders with IT and business stakeholders across different departments.

4. **Build expertise:** Develop an internal D&A culture and capabilities as your first guardian of trust. D&A practitioners are critical to being able to elevate the wider understanding of D&A across the organization. Identify gaps and opportunities in your current capabilities, governance, structure and processes. Ensure that you have expertise in analytics quality assurance: experimental design, A|B testing, and other means of validation. Ultimately, make trust in D&A a core company value.

5. **Encourage transparency:** Open the "black box" to a second set of eyes. And a third. There are many potential actions to help improve D&A transparency. You may want to establish cross-functional teams, third-party assurance and peer reviews, use wiki-style sites, encourage whistleblowers and strengthen QA processes as valuable "guardians" of trust. Essentially, have every D&A challenge reviewed independently.

6. **Take a 360-degree view:** Build your ecosystems, portfolios, and communities. To drive trust through the organization, you will need to look beyond the traditional boundaries of systems, organizational silos, and business cases to see the wider ecosystems. Take a portfolio approach, looking at the value and the risk that D&A brings to the organization. Create a "meta-model" and cross-functional teams to identify and control dependencies between models.

7. **Be innovative—Enable experimentation:** Create a model for D&A innovation. Allow D&A teams to push the boundaries of innovation and try several paths without excessive fear of failure. Build a data innovation lab,

which allows data scientists and business stakeholders to rapidly test new ideas. Consider ROI beyond the specific performance objectives of the D&A project. Find ways to incentivize employees for innovation and trusted D&A.

Trust is a central component of successful D&A programs, which ultimately inform some of the most important decisions that businesses make. Organizations that are unwilling or unable to trust their D&A will fall behind and be considered laggards in the application of what is soon to be a ubiquitous agent of change and accepted standard of practice. Those that can implement trusted analytics and act confidently on their findings will have the greatest advantage in the corporate world.

Chapter 4

The Missing Link: Experiential Learning

Natalia V. Smirnova and Lorri A. Halverson

Contents

Introduction

This chapter is a practical guide for implementation of a collaborative, experiential learning exercise. This program recognizes a connection between the concept of intuition and experience, and the importance of building these competencies before entering the workplace as well as throughout one's career. The literature surveyed genuinely aided an understanding of the experiential learning concept, which then helped to design the experiential learning program.

Practical experience in the form of experiential learning fosters the development of intuition and, more specifically, an intuitive leader and a successful executive. By looking at experiential learning through two lenses, the intuition and management literature as well as the literature about engaged learning and pedagogy of experience, it can be shown that experiential learning is the missing link in the development of a skilled workforce with a high degree of executive function. Specifically, the argument is that experiential learning opportunities, which are set up within collaborative arrangements between academia and practitioners, contribute to the development of three things: core knowledge, skills to apply such knowledge, and intuition. We use the notion of intuition defined by Herbert A. Simon (1992) as a recognition of experiences "encapsulated" in memory.* This allows us to argue that, as students move into the workplace they gain more expertise and strengthen the knowledge base from which they will develop intuition. This includes both intuitive judgment and intuitive insight, a distinction established by Dörfler and Ackermann (2012). The research on the development of intuition, and ultimately its use in executive decision-making, is based on the assumption that these decision makers are experts in their professional fields. That is, expertise precedes intuition. As such, experiential learning provides a link between the structured, theoretical knowledge base of the classroom and the messy, "ill-structured" problems of the workplace, establishing the basis of expertise. As expertise grows and matures, one recognizes more and more cues from past experiences, therefore developing deeper intuitive knowledge, intuitive insight, and confidence. These ingredients provide the underpinnings of the executive decision-making process and, therefore, support the development of the intuitive leaders especially valuable in the twenty-first century workplace.

The changing nature of the workplace is documented by many academic research papers and popular articles. For example, see Newman and Winston (2016), Christensen and Schneider (2010), Tripathi and Sharma (2011), and Trilling and Fadel (2009), among others. It is common knowledge that many people around the world work remotely, telecommute, and teleconference. Working in teams, and particularly in teams of people with diverse sets of skills, backgrounds, and cultural competencies, will become even more prevalent in the coming years. These labor market changes, as well as the economy's continued transformation from manufacturing to service, global competition, and the need for constant retooling and learning new skills call for alterations in the ways students are prepared for future careers.

Considering these changes, experiential learning can bridge classroom knowledge with the workplace by laying a foundation for the development of intuition and expertise. There are many examples of experiential learning depicted in theoretical and practitioner's literature. Examples such as apprenticeships and internships have received much attention. In this chapter, an example of experiential learning is described. It is a collaborative program between academic institutions and the practitioners aimed at advancing content knowledge through immersion in

---

* Simon (1992), p. 156.

real-world projects. Participating students experience the workplace environment, in all of its challenges, learn soft skills needed to navigate that environment, and establish a base of experiences on which they can build. Supported by qualitative evidence from students' reflections, we are confident that the development of those traits resulted from a structured experiential learning program.

The applied research course that we are showcasing in this chapter is an application of the experiential learning model, which blends the academic and practitioner's sides of gaining expertise and intuition. As such, we argue that this course could be viewed as the missing link in developing executive leaders. In the next section, a survey of the literature about intuition and practical experience illustrates the theoretical base for this application. From there, this application is described in more detail and the transmission mechanism of the "missing link" is shown. In the final section, readers are invited to follow the step-by-step guide for replicating this innovative approach and developing their own collaborations.

Intuition and Practical Experience: The Literature

This section summarizes the existing consensus in the scientific community regarding various aspects of experiential learning. The theoretical underpinnings of the following features are showcased: intuition and its links to expertise and decision-making; methods of student engagement outside of the classroom; core skills of the contemporary workplace; specificity of economics; and partnership building. These features guided us in building our applied economic research course.

Intuition and Its Links to Expertise and Decision-Making

Simon (1992) defines intuition as simply a recognition of the cues stored in the memory of an expert. As more experience is gained by an individual, the more information is stored in her memory. As a situation presents itself, the cue is provided, and needed information is retrieved. We use this definition of intuition as a basis for our experiential learning application.

Dörfler and Ackermann (2012) argue that two distinguishable forms of intuition exist: intuitive insight and intuitive judgment. In doing so, the authors identified several dimensions of intuition's typology. With the intention to connect intuition to experiential learning and ultimately executive decision-making, the focus is on understanding intuition types that relate to solving real-world problems. Dörfler and Ackerman (2012) suggest that intuitive judgment manifests itself in "deciding about an alternative or about a direction," while intuitive insight is concerned with "creating a solution which entails new knowledge."* We claim that the development of both types is important for an executive leader.

---

* Dörfler & Ackermann (2012), p. 559.

When looking at traditional classroom functions including problem sets, case studies, research projects, and so on, an argument can be made that this only develops the student's intuitive judgment. However, experiential learning opportunities contribute to both well-structured and ill-structured projects for students to tackle and to build an intuitive insight. We built our applied economic research course to combine core knowledge gained in classroom learning with experiences, therefore contributing to the advancement of both: intuitive judgment and intuitive insight. If such ambitious outcomes are to be realized, the students who experience the course will be well-positioned to enter the workforce with the basis for becoming a leader in an organization.

The literature in a variety of disciplines distinguishes between two systems of cognition: System 1 that is fast, holistic, intuitive; and System 2 that is slower, more analytic, and cognitively effortful (Kahneman, 2011). We argue that since System 2 functions in an abstract, sequential, and rule-based manner, it can be developed through the academic exercises of sense making, mental simulation, situation assessment, and problem representation. System 1 uses functions in a domain-specific manner and, therefore, is developed or enhanced through experience in the work setting or through applied work in general.

The role of intuition in strategic decision-making is explored by Sinclair, Eugene, and Hodgkinson (2009). The authors called for the development of programs that educated business leaders in the complimentary use of both intuition and analysis. They found those programs to be valuable in fostering efficient decision-making processes. In response to their challenges, our pilot program was created focusing on core knowledge, application of this knowledge, and evolving intuition, specifically paying attention to System 1 development. By evaluating the results of the pilot, we rolled out the program with a greater emphasis on leadership skills acquisition.

Betsch (2008) specifies that knowledge acquired through experience is the input into intuitive processing. Salas, Rosen, and DiazGranados (2010) state that the expertise-based intuition is based on combining the two domains: expertise and intuition. Kahneman (2011) argues that expert intuition is a skill and needs three things to develop: stable environment, practice, and feedback. These authors promote associative learning, which is understood here to be synonymous with engaged, applied, experiential learning. This learning, which occurs outside of the classroom, is a hands-on, real-world immersion into the workplace environment.

Methods of Student Engagement Outside the Classroom

In addition to compelling evidence from the literature that learning outside of the academy develops expert intuition, evidence from best practices of internships, apprenticeships, and other methods of engagement of students outside of the classroom was examined. As the applied research course application was established, the ideas of Kolb (1984) were explored and the contemporary views of them, summarized by Moore (2013), was heavily drawn upon. The three pillars of the experiential learning approach are in the foundation of our approach. The first pillar is *Learning from Experience* introduced by Dewey (1938). This theory was applied by involving students in the professional life of the practitioner. The research assignments that were given dealt not only with economic content, but also included workflow schedules, production deadlines, staff reassignments, and other naturally occurring workplace events.

The second pillar of the experiential learning approach is *Communities of Practice* introduced by Lave and Wenger (1991) and expanded by Wenger (1998). This idea postulates that learning is a function of active participation in the changing environment of the workplace. With staff researchers already forming a community of practice, students were organically integrated into it. This arrangement allowed them to experience the community of practice and engage in professional interactions within it.

The third pillar is *Workplace Learning* described by Billet (2001), Fenwick (2003), and Raelin (2008). This pillar suggests that each workplace has certain intrinsic characteristics that are specific to it, and as such, employees adjust to them. While the learning that occurred during the program was specific to the practitioner's organizational structure and research agenda, care was taken to guide students through reflective exercises at the end of the course. The goal was to help them articulate the skills and competencies they acquired during this experience and understand which proficiencies are organization-specific and which ones are universal.

Core Skills of Contemporary Workplace

What about the core skills, as opposed to intuitive insights, that are being acquired in the workplace? Are they different from the ones that can be attained in the classroom? Moore (2013) posits that the learning that occurs in the workplace is not simply an "application" of the concepts acquired in the classroom. The out-of-school environment requires different kinds of cognition, and thus the skills that are attained at the professional site are completely different from the skills developed in the academic setting. A portion of the value of practical experience is, then, in the additional cognitive abilities being developed; two of them, we argue, are expertise and intuition.

The support of new learning goals for the workplace environment in the twenty-first century is presented in Newman and Winston (2016) and Trilling and Fadel (2009). Both books argue that because the dynamics of the workplace call for teamwork, flexibility, and situated learning, the skills that are required in the practical world are unlike the skills and competencies attained in the classroom. Trilling and Fadel (2009) suggested that there are three groups of skills that are necessary for functioning well in the twenty-first century: (1) learning and innovation; (2) digital literacy; and (3) career and life skills. These skills are part of a unified, collective vision for twenty-first century learning that was developed by the Partnership for 21st Century Learning (n.d.).

For learning and innovation skills, the literature specifically mentions critical thinking and problem solving, communication and collaboration, and creativity and innovation. In an experience-based education, the practice of critical thinking assumes a form different from the academic setting. Critical thinking at the workplace advocates action, preparation of well-articulated ideas, and efficient delivery of those ideas to others.

For the digital literacy skills, Trilling and Fadel (2009) discussed information literacy, media literacy, and information and communication technology (ICT) literacy. The increased demand for the ability to access information efficiently and effectively, evaluate information critically and competently, and use information accurately and creatively. Media literacy is described as the ability of students and employees to competently operate in various delivery methods (print, graphics, animation, audio, video, websites, etc.). The ICT literacy calls for the ability to use technology as a tool to research, organize, evaluate, and communicate information, as well as use digital technologies appropriately and ethically. By eliminating the classroom "walls," students are exposed to the current technology tools and experience the challenges and limitations inherent to them.

For career and life skills, the literature points to flexibility and the ability to adapt, initiative and self-direction, social and cross-cultural interaction, productivity and accountability, leadership and responsibility. While those are all inherent to the work environment, engagement in real-world projects also develops the concept of self and identity, for example, an understanding of what it means to "be" an economist.

Moore (2013) argues that under the right conditions, the two forms of knowledge (academic and experiential) could be compatible, complementary, and mutually expansive. The potential for the workplace adding value to the school's curriculum and vice versa is a matter of pedagogy, teaching practices, and institutional missions.*

Specificity of Economics as a Field of Study

The cognition, intuition, and core skills acquired in the workplace appear to be universal. That is, they can apply to any field of study. To determine if something distinct happens in economics, in terms of students' engagement, the economic education literature was reviewed. The seminal paper by Allgood, Walstad, and Siegfried (2015) summarized the methods of teaching economics to undergraduate students, and provided an opportunity to assess which methods are proven to be most beneficial to students in the academy. For example, cooperative learning, experiments, classroom and online discussions are all identified as the pedagogies that stimulate engaged learning and provide positive outcomes. Even though Allgood et al. (2015) do not provide recommendations for the collaboration of

---

* Moore (2013), p. 99.

academic institutions with practitioners in the field, the conclusion that engaged learning is the most beneficial to student achievement is encouraging.

In addition to engaged learning, the assessment of innovative structures of capstone courses in economics provided the foundation for our approach. For example, Li and Simonson (2016) describe the setup and the outcomes of a senior research course in economics in a public university. They concluded that the redesign of the course to provide an overview of the scientific research process showed positive effects on student learning outcomes and on their satisfaction with the course. The model presented in this chapter builds on this idea by involving students in the actual scientific research process.

Partnership Building

Last, the literature pertaining to the creation of collaborative partnerships between academic institutions and practitioners was surveyed. Fry and Kolb (1979), for example, suggest that the experiential learning approach integrates perceptive understanding of concepts with practical experience and is to be used in liberal arts colleges. In subsequent decades, many partnerships were formed, and the literature regarding them is abundant. Reardon (2000) describes the value of a university's community engagements and Ishisaka, Sohng, Farwell, and Uehara (2004) present an example of an "engaged" social work education. Gavigan (2010) showcases the value of integrating career offices on campus with the academic departments to provide meaningful summer internships opportunities for students.

The experiential learning model that is the focus of this paper went even further. While the collaboration between the academy and practitioners resembles the projects described in the above-cited papers, the scope of students' work is different. The uniqueness of this project is that students were engaged in the practices specific to their future professional field (in this case, economics). This expansion allowed students to not only contribute to the output of the organization, but to also gain an understanding of what it means to be an economist.

With the foregoing research as a base, a new experiential learning application was conceived. The main goal of this project was to bridge the two knowledge domains in which students operate: the academic classroom and the workplace. By using the theoretical underpinnings of intuition development, engaged learning, and the curriculum of experience, and by applying experiential pedagogy, the university classroom and the professional environment were integrated.

Since the American Institute for Economic Research (AIER) is an economic think tank, it is focused on providing value-added programs to students studying or intending to study economics, as well as those studying related disciplines such as business, finance, accounting, or entrepreneurship. Thus, a natural partner from academia was the Business School of the University of Sioux Falls. Participating students were from diverse courses of study: economics, marketing, accounting, and business. During the project, every effort was made to devise opportunities for students to debrief in ways that highlighted the possible synergies between the academy and the real world, as well as the development of knowledge, competencies, skills, and expert intuition relevant to their future careers. It is therefore argued that an applied research course in any discipline is a vehicle to provide the missing link from core knowledge to expertise and intuition.

The next section ties together the experiential learning partnership model with the goal of developing highly effective leaders who can employ intuition in the executive arts. Based on the collected qualitative evidence of improved students learning outcomes, a suggested process by which similar experiential learning partnerships can be built has also been included.

The Experiential Learning Partnership: The Missing Link

The focus of this section is to argue that the integration of academia and the workplace is beneficial to all parties involved and that those benefits spill over to firms seeking students as potential employees with this expanded skill set. Building expert knowledge and intuition through experiential learning contributes to the development of future leaders of the profession. It benefits the students as well as their future employers because they carry with them management experience and people skills. Though the case described here relates to economics, there is a high degree of confidence that this approach can be replicated in any industry.

The core idea and related applications are based on the education literature that postulates that the experiential learning in the workplace is not just the opportunity to "apply" the material learned in the classroom, but it complements the academic learning by developing life skills and intuition that are highly transferrable to any environment.

Figure 4.1 shows that the applied economic research course, described here, encompasses the core knowledge gained in the classroom and intuitive insight obtained in the office setting. Therefore, it is hypothesized that this experiential

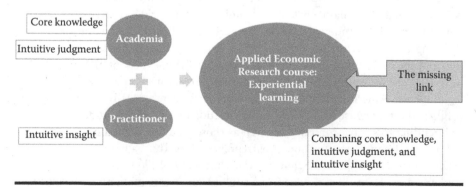

Figure 4.1 The missing link: Experiential learning.

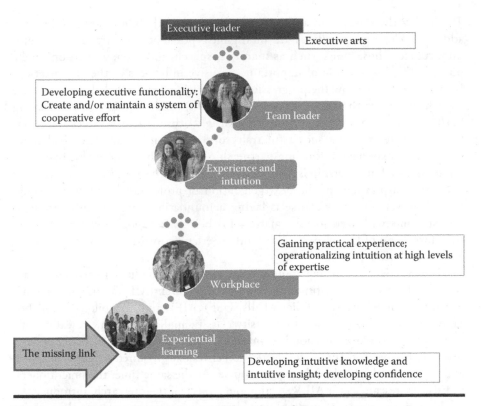

Figure 4.2 Skills and intuition acquisition throughout a career.

learning opportunity connects young people on a career path with a leadership track. The contribution of this approach is the recognition that experiential learning creates the potential for students to become executive leaders.

The progression of students' skills and intuition acquisition is shown throughout a career in Figure 4.2. The starting point of the trajectory is an applied research course conducted in the form of experiential learning. Upon graduation, using intuitive knowledge, intuitive insight, and confidence gained in the course, he or she enters a workplace, gaining additional practical experience and operationalizing intuition. With experience and intuition in hand, an expert develops into a team leader and eventually into an executive. The executive functions serve to create and maintain a system of cooperative effort.* The operation of such systems requires the highest development of the executive arts.†

To clarify this transmission mechanism, we refer to the classic book of Barnard (1945) on the functions of the executive. In this treatise, Barnard argues that the development of leadership abilities is of most importance for an executive.

---

* Barnard (1945), p. 216.
† Ibid, p. 222.

The higher the position of authority, the more general abilities are required. In addition to leadership abilities, the intangible types of personal traits are of major importance. Those traits, such as manners, speech, intuition, and so on, make possible the development of important executive influences in the organization. In Figure 4.2, we show the progression of acquiring core knowledge, experience (which embodies the application of those skills), and intuition (encompassing both: intuitive knowledge and intuitive insight) throughout a career. The contribution here is that a successful trajectory of becoming an executive leader starts at the experiential learning partnership model. Throughout this paper, we argue that such partnership is a missing link between the educational attainment and career aspirations of the next generation of professionals. We fully recognize, however, that this course is laying a foundation for such an alignment. The students will need to take matters in their own hands to ensure that the transformation of knowledge, skills, and experiences into an executive functionality does materialize.

The experiential learning application is now being conducted between a practitioner, the American Institute for Economic Research (AIER), and an academic institution, the University of Sioux Falls (USF). AIER and USF piloted a collaboration to seamlessly integrate a university class in applied economic research with the ongoing work of an economic think-tank. The students register for the course, meet at the specified times, complete the deliverables required by the instructor, and are assigned a final grade for the course. At the same time, the practitioners, in this case researchers at AIER, section out a piece of their research, introduce the students to the research question, provide mentoring to the students throughout their research, and offer robust feedback like what would be received if the students were the practitioner's employees.

While the daily work of the students is done under the tutelage of the faculty member, there exists an expert researcher who is to guide, support, and further challenge the students in this endeavor. To minimize the supervisory burden on the practitioner, students are assembled into teams which tackle different projects that are within the research agenda of the practitioner. To sustain this collaboration, students engage with their assigned researchers via videoconference during pre-established times in the course, much in the same way business colleagues in locations around the globe would meet electronically to discuss a shared project. During these meetings, the mentor-researchers provide feedback, answer questions, and further develop personal relationships important to any team endeavor.

In terms of the students learning outcomes, we identified gains in both hard and soft skills. Students expanded hard skills of economic research such as developing research proposals, designing a research plan, compiling data, and writing a complete academic paper. In terms of soft skills, identified in Heckman and Kautz (2012) and elaborated by Deming (2015), students reported that through this program they learned to be confident, independent thinkers. They learned to self-guide their research progress, and communicate and defend their arguments and results to the experts.

The debriefings with students during and after the completion of the course were done robustly and with great care. Kahneman (2011) argues that timely feedback is essential to the development of expert intuition. During the regularly scheduled conversations, the professor has an opportunity to mentor students, correct courses of action, and help eliminate roadblocks. During the final debriefings with practitioners, students receive constructive, corrective feedback confirming their perception of the work environment, research processes and products. During those conversations and in formal written evaluation forms, students identified their confidence in being able to enter and grow in the profession of their choosing. They were enthusiastic about entering the workforce and becoming successful practitioners themselves.

Students' reflections, presented in Table 4.1, support the positive learning outcomes that we set as goals for this program. To meet the demands of the twenty-first century, students become knowledgeable, adaptable people who can work with others to innovate and thrive in the new economy.

Table 4.1 Program Outcomes

Outcomes	Qualitative Evidence
Research Process	"I learned what data analysis looks like and what it takes to do it."
	"I know not to overlook the work that goes into finding this kind of data."
	"Listening to the supervisors' presentations gave me a lot of new knowledge and allowed me to analyze different research in ways I hadn't thought of doing before."
Communication	"Focus on tasks, focus on the outcomes, and convey the big picture. Don't get bogged down by little details in a presentation."
	"Learning to communicate more openly is something I believe I can relate to any job to make sure that the work I am doing is accurate."
	"What I realized was that I am capable enough to stand in front of people with PhDs and speak in a sensible way."

(Continued)

Table 4.1 (*Continued*)

Outcomes	Qualitative Evidence
Professionalism	"I gained a better understanding of the professionalism that it takes to excel in the business world and the importance of conducting yourself in a professional manner at all times."
	"I gained experience working with professionals in the workplace."
	"I learned that economic researchers are not just anti-social people who lock themselves in offices doing work all day. … These researchers are smart, intelligent people who are problem solvers, statisticians, and people with personality who appreciate sarcasm (which I thoroughly enjoyed)."
	"AIER has given me an opportunity to partake in an economic research environment, my first experience outside of the services industry."

How to Replicate This Approach

Our vision is to build a network of academia-practitioner collaboratives throughout the country, and, with the help of technology, throughout the world. While the model has been developed, and tested with a focus on economic research, the structure of the model and ultimately, the positive results, are not limited to that field. Practitioners can be pulled from policy think tanks, private industries, national nonprofits, and government agencies, to name just a few.

To make it easier for aspirants in these fields and others to replicate this program, the implementation has been distilled into a 10-step process. These steps reflect the learning from the initial collaboration and subsequent iterations including the expansion of the model to a new partner university. Potential collaborators can step through the process as it is summarized in Table 4.2.

In the second iteration of this program, another academic institution—Missouri University of Science and Technology—joined USF and AIER. This experience allowed for the identification of a first-year learning curve. Both, the "new" professor and the AIER supervisor needed to adjust their expectations about what could feasibly be accomplished, what the dynamics of the teams would be, and how to solve scheduling and technology related problems. Even though one could theoretically follow the 10-step guide (Table 4.2), the implementation of each step will take time and effort. However, if the academic and the practitioner are prepared to be flexible and attuned to each other's needs, the program will still be a success. The

Table 4.2 Experiential Learning Partnership Model—Building a New Partnership

Steps	Description
Step 1: **Develop a Partnership**	The partnership is the bedrock of this model and requires a significant infusion of time for both the university and the practitioner. It should be understood from the start that full implementation of this model will cover more than one academic year and may require additional investments in technology and staff time.
Step 2: **Select or Develop the Course**	This model for experiential learning needs to relate to a class where the students have committed time and are invested in the outcome. For the piloted program, with its focus on economic research, the course needs to include prerequisite courses in microeconomics, macroeconomics, and statistics. It is the belief of the authors that this model will work successfully in other disciplines as well. In those cases, modifications will need to be made to ensure students enter the class with the requisite knowledge and skill set to complete the project to the level required by the practitioner. Without those skills, the benefits of the experiential learning model break down.
Step 3: **Establish Desired Deliverables and Outcomes**	While the precise learning outcomes and deliverables need to be established by the university partner, it must meet the requirements of both parties and there needs to be clarity around these issues before the course begins. In the initial pilot, it was established that students would produce an academic paper in a format that was ready for publication. When the program was expanded, this question needed to be addressed by the new university. In this case, less emphasis was placed on writing and a greater focus was given to the data acquisition, analysis, and presentation of results. Both sets of outcomes and deliverables functioned well within the model and still use experiential learning to link core academic knowledge with the ongoing work of the practitioner.

(Continued)

Table 4.2 (*Continued*)

Steps	Description
Step 4: Enroll Students	While this is intuitive to most, there are additional considerations that need to be attended to at the university level including: registration timeline, the role of the course in the students' degree plan, and enrollment management. One challenge worthy of note is the inherent mismatch on timelines between a university and a practitioner. That is, the enrollment process happens very far in advance of the course. Given the fluid nature of the projects being proposed by the practitioner, students may need to register for the course without knowing the exact focus of the research or project.
Step 5: Select a Topic	The selection of the final research topic falls heavily on the practitioner. Moreover, to make this process meaningful for both partners, it should contribute to the work being done by the practitioner. As the topics are developed, practitioners need to be aware of staffing strengths and limitations. While this process is often an enjoyable change for the researchers and other employees who provide the mentoring and feedback, it can also be quite time consuming and draw them away from other work. Practitioners will also need to anticipate potential staff changes and how they will fill the resulting void. An additional consideration for the practitioner will be selecting topics or projects that allow the practitioner employees to showcase their interests and passions. It has been observed that on the topics where the mentoring researchers were highly interested and engaged, the students were also more interested and engaged. This all requires a high level of flexibility on the part of the university faculty member. Specifically, it may be the case that the topics and/or project details are not available until immediately before the course begins.
Step 6: Develop a Work Plan	A challenge of working within the confines of a semester-long course is that the work plan will need to incorporate the academic calendar as well as time for practitioner to provide meaningful feedback. Having pre-scheduled touch points between the students and their mentor researchers allowed students and researchers to plan their schedules to ensure full participation.

Table 4.2 (*Continued*)

Steps	Description
Step 7: **Assign Students to Teams**	One of the successes of the pilot program was the ability to match students together in balanced teams that included a mix of quantitative, technical, and writing skills. The noticeable imbalance between teams is an example of the "ill-structured project" and as such it supports the learning goals of this model.
Step 8: **Coordinate Program Details**	During the semester, the mentoring researchers and the students came together via WebEx. These videoconferences were important to introduce students and mentors, to establish trust, and to provide a forum for ideas, questions, and concerns. Coordination extended well beyond the setting of dates and times. Resources are required on both ends to ensure that the proper technology is in place and working. Additional coordination is required to track team progress and meet assessment requirements.
Step 9: **Assess Program Successes and Shortcomings**	Ultimately, the model exists to produce results. Considering that, short-term and longer-term student assessments are needed to measure the degree to which the experiences of the students are sufficiently advancing their level of professional expertise and supporting the development of intuition. This has been done with open-ended reflection questions, online surveys, and structured debriefing sessions immediately upon completion of the course. Longer-term follow-up methods and measurements are still being developed. The feedback from the students will be combined with the employer survey data to match the experiences in this learning model with the expertise needed in various industries.
Step 10: **Modify the Program**	As the research reflects, the workplace continues to change and, as such, ongoing adjustments will need to be made to the model. The immediate feedback from students allows for adjustments to be made from one academic year to the next. Incorporation of longer-term feedback will, naturally, positively change the model in more subtle ways.

students were generally spared from encountering the hiccups around logistics, but the professor must be prepared to manage these in a timely manner.

To enact the vision is to scale up in a way that will include potentially many more partners in a single collaboration. It is the belief of the authors that the work of the practitioners can, and ultimately will, be leveraged across several academic institutions and many students. As this collaboration grows, stacking the process of joining the group semester-by-semester is highly recommended. That is, to minimize the impact of the first-year learning curve mentioned earlier, it is recommended that only one or two new partners be added with each new iteration. This will allow the practitioners to work more closely with the new institutions and curtail the disruptions for the other partners. This way, after the first iteration of the program, in each semester the "new" entrant will learn the ropes, while the "experienced" institutions will pull the project forward. We recommend, initially, that the teams be comprised of students from one university. Only later, when all parties are seasoned participants, the practice could be broadened to mix students from different universities to serve on the same team. This experience will enhance students' confidence, ability to resolve conflicts, and expert intuition.

Conclusion

As the literature documents, changes in the labor market, combined with global competition, require a change in how human capital is developed. As such, by breaking down the walls between college classrooms and the institutions that will employ college graduates, we expose students to the workplace practices and actively engage them in the learning process. The hands-on nature of our project forced students to be flexible as they experienced challenges working in groups, adjustments to the scope of their projects, or changes in the research staff supporting their work.

It is understood that the skills needed in young graduates today are not likely to be the skills that will carry them for their entire careers. Partnerships, such as the one described in this chapter, provide opportunities for problem solving, communication, and collaboration, and develop a foundation for skills that are transferable. Such partnerships are to be fostered. Beyond the application of economic theories, students gained an understanding of research methodologies, proficiency in digital literacy, and advanced skills in data management and analysis.

The contribution of the experiential learning exercise showcased in this chapter is twofold. First, is the development of the 10-step guide on how to build a new partnership. This guide provides a starting point for new university-practitioner partnerships in various fields of study. The second contribution is the illustration of the theoretical transmission mechanism from experiential learning through intuition to executive arts. It is the goal of future studies to empirically support such a claim. We want to caution, however, that the availability of such experiential learning partnerships and the transmission mechanism does not guarantee the conversion to a successful executive leader for every person. It is up to each individual student who goes through the program to take this opportunity and develop it to fruition.

The students in university classrooms today are the employees and leaders of the next decade. By investing time in developing partnerships between academic institutions and practitioners, we are integrating the academy and the workplace, broadening the base of experiences on which potential employees will draw, and striving to enhance the intuitive decision-making of future executives.

References

Allgood, S., Walstad, W. B., & Siegfried, J. J. (2015). Research on teaching economics to undergraduates. *Journal of Economic Literature, 53*(2), 285–325.

Barnard, C. I. (1945). *The functions of the executive*. Cambridge, MA: Harvard University Press.

Betsch, T. (2008). The nature of intuition and its neglect in research on judgement and decision making. In H. Plessner, C. Betsch, & T. Betsch (Eds.), *Intuition in judgement and decision making* (pp. 3–22). Mahweh, NJ: Lawrence Erlbaum.

Billet, S. (2001). *Learning in the workplace: Strategies for effective practice*. Crow's Nest: Allen and Unwin.

Christensen, K., & Schneider, B. (Eds.). (2010). *Workplace flexibility: Realigning 20th-century jobs for a 21st-century workforce*. Ithaca, NY: ILR Press. EconLit with Full Text, EBSCOhost.

Deming, D. J. (2015). *The Growing Importance of Social Skills in the Labor Market*. NBER Working Paper No. 21473. Retrieved from http://scholar.harvard.edu/files/ddeming/files/deming_socialskills_aug16.pdf (Revised from 2016, August).

Dewey, J. (1938). *Experience and education*. New York, NY: The Macmillan Company.

Dörfler, V., & Ackermann, F. (2012). Understanding intuition: The case for two forms of intuition. *Management Learning, 43*(5), 545–564.

Fenwick, T. J. (2003). *Learning through experience: Troubling orthodoxies and intersecting questions*. Malabar, FL: Krieger Publication.

Fry, R., & Kolb, D. (1979). *New directions for experiential learning, Enriching liberal arts through experiential learning* (pp. 79–92). San Francisco, CA: Jossey-Bass.

Gavigan, L. (2010). Connecting the classroom with real-world experiences through summer internships. *Peer Review, 12*(4), 15–19.

Heckman, J. J., & Kautz, T. (2012). Hard evidence on soft skills. *Labour Economics, 19*(4), 451–464.

Ishisaka, H. A., Sohng, S. S. L., Farwell, N., & Uehara, E. S. (2004). Partnership for integrated community-based learning: A social work community-campus collaboration. *Journal of Social Work Education, 40*(2), 321–336.

Kahneman, D. (2011). *Thinking, fast and slow*. New York, NY: Farrar, Straus and Giroux.

Kolb, D. A. (1984). *Experiential learning*. Englewood Cliffs, NJ: Prentice Hall.

Lave, J., & Wenger, E. (1991). *Situated learning: Legitimate peripheral participation*. New York, NY: Cambridge University Press.

Li, I., & Simonson, R. (2016). Capstone senior research course in economics. *Journal of Economic Education, 47*(2), 161–167. doi:10.1080/00220485.2016.1146103

Moore, D. T. (2013). *Engaged learning in the academy: Challenges and possibilities.* New York, NY: Palgrave MacMillan.

Newman, K. S., & Winston, H. (2016). *Reskilling America: Learning to labor in the twenty-first century.* New York, NY: Metropolitan Books.

Trilling, B., & Fadel, C. (2009). *21st Century skills: Learning for life in our times.* San Francisco, CA: Jossey-Bass.

Tripathi, V., & Sharma, A. (2011). Unfolding cross cultural issues at work place: A real HR challenge. *Indian Journal of Social Development, 11*(1), 79–89. EconLit with Full Text, EBSCOhost.

Partnership for 21st Century Learning. (n.d.). *Framework for 21st Century Learning.* Retrieved from http://www.p21.org/our-work/p21-framework (Accessed 2016, May 17).

Raelin, J. A. (2008). *Work-based learning: Bridging knowledge and action in the workplace* (New and revised ed.). San Francisco, CA: Jossey-Bass.

Reardon, K. M. (2000). An experiential approach to creating an effective community-university partnership. *Cityscape: A Journal of Policy Development and Research, 5*(1), 59–74.

Salas, E., Rosen, M. A., & DiazGranados, D. (2010, July). Expertise-based intuition and decision making in organizations. *Journal of Management, 36*(4), 941–973.

Simon, H. A. (1992, May). What is an "explanation" of behavior? *Psychological Science, 3*(3), 150–161.

Sinclair, M., Eugene, S.-S., & Hodgkinson, G. P. (2009). The role of intuition in strategic decision making. In L. A. Costanzo, & R. B. MacKay (Eds.), *Handbook of research on strategy and foresight* (pp. 393–417). Northampton, MA: Edward Elgar.

Wenger, E. (1998). *Communities of practice: Learning, meaning, and identity.* New York, NY: Cambridge University Press.

TRUST

II

Chapter 5

The Foundations of Trust

Joanna Paliszkiewicz

Contents

Introduction

Nowadays, in the era of different crises, the role of trust is very important. Trust is difficult to establish and sustain. It is also easy to destroy. Trust is very fragile.

Many researchers and practitioners have explored the significance of trust in the organization (Bibb & Kourdi, 2004; Covey, 2009; Lewicki & Bunker, 1996; Mayer, Davis, & Schoorman, 1995; Paliszkiewicz, 2013; Paliszkiewicz & Koohang, 2016; Six, 2004; Sprenger, 2004).

According to researchers, building trust is essential to establishing employee loyalty (Costigan, Ilter, & Berman, 1998), commitment (Eikeland, 2015; Lewicka & Krot, 2015), communication and information exchange (Doney & Canon, 1997; Hakanen & Soudunsaari, 2012; Li, Poppo, & Zhou, 2010; Malhotra & Murnighan, 2002; Morgan & Hunt, 1994; Tyler, 2003; Young & Daniel, 2003), and effectiveness and organizational performance (Lewis & Weigert, 1985; McAllister, 1995; Nooteboom, 2002; Paliszkiewicz, Gołuchowski, & Koohang, 2015; Paliszkiewicz, Koohang, & Horn Nord, 2014; Pangil & Chan, 2014). Trust also mitigates uncertainty about partner behavior (Krishnan, Martin, & Noorderhaven, 2006).

The dynamics of trust enhancement are a very important topic of research (Muethel & Bond, 2013).

The aim of this chapter is to present the foundations of trust based on a critical literature review and to propose future research in this area. In the first part of the chapter, the definitions of trust are presented. Next, the types of trust and the methods for building trust are described. Finally, the practical implications, conclusions, and future directions are proposed.

Trust Definitions

The concept of trust has been researched in various disciplines, such as psychology, sociology, economy, management, marketing, ergonomics, human-computer interaction, and electronic commerce. However, researchers from different fields take different approaches to understand this phenomenon (Yoon, 2014). There is a consensus among scholars that trust is difficult to define. It is easier to define distrust than trust. However, several efforts to derive a definition of trust from different disciplines have been made. The most common definitions of trust are listed in chronological order in Table 5.1.

Trust in the literature has also been defined as a probabilistic evaluation of trustworthiness, as a relationship based on ethical norms or as an agent's attitude (Taddeo, 2009). Psychologists underline that trust is a personal trait. Per sociologists, without trust, you cannot create any relationship and the trust is voluntary. In business, trust usually forms as part of a formal contract. In management, it is an employee's loyalty to execute an organization's goals (Paliszkiewicz & Koohang, 2016).

According to Paliszkiewicz (2013), trust is the belief that another party: (1) will not act in a way that is harmful to the trusting firm, (2) will act in such a way that it is beneficial to the trusting firm, (3) will act reliably, and (4) will behave or respond in a predictable and mutually acceptable manner.

Trust can be described as interpersonal and it cannot be compulsory. It appears between two individuals and depends on the situation. Trust is dynamic, very easy to destroy, and difficult to rebuild. Sometimes it is even impossible to rebuild trust.

Types of Trust

Trust is multidimensional and complex, which is why there are so many propositions in the literature for different types of trust. The most common types of trust are listed in chronological order in Table 5.2.

Table 5.1 The Definition of Trust

S.N.	Authors	Definition of Trust
1.	Wrightsman (1966) Rotter (1967)	Trust is a personal trait, which is responsible for the general expectancies of the trustworthiness of others.
2.	Gibb (1978)	Trust is instinctive, and, as a feeling, is so close to love.
3.	Sako (1992)	Trust can be treated as a state of mind, an expectation held by one partner about another, that the other behaves or responds in a predictable and mutually acceptable manner.
4.	Morgan & Hunt (1994)	Trust is one party's confidence in an exchange partner's reliability and integrity.
5.	Lewicki & Bunker (1995)	Trust is a state involving confident positive expectations about another's motives regarding oneself in situations of risk.
6.	Mayer et al. (1995)	Trust is the willingness of a party to be vulnerable to the actions of another party based on the expectation that the other will perform an action important to the trustor, irrespective of the ability to monitor or control that other party.
7.	McAllister (1995)	Trust is a cognitive judgment about another's competence or reliability and an emotional bond of an individual toward the other person.
8.	Bidault & Jarillo (1997)	Trust is believing that the other party will behave in our best interests.
9.	Doney & Cannon (1997)	Trust is a willingness to rely on another.
10.	Gambetta (1998)	Trust (or, symmetrically, distrust) is a level of the subjective probability with which an agent assesses that another agent or group of agents will perform an action.
11.	Bhattacharya, Devinney, & Pillutla (1998)	Trust is an expectancy of positive (nonnegative) outcomes that one can receive based on the expected action of another party in an interaction characterized by uncertainty.

(Continued)

Table 5.1   The Definition of Trust (*Continued*)

S.N.	Authors	Definition of Trust
12.	Rousseau, Sitkin, Burt, & Camerer (1998)	Trust is a psychological state comprising the intention to accept vulnerability based on positive expectations of the intentions or behavior of another.
13.	Sztompka (1999)	Trust is the expectation that other people, groups, or institutions with whom we interact will act in ways conducive to our well-being.
14.	Tatham & Kovacs (2010)	Trust as a nonrational choice of a person faced with an uncertain event in which the expected loss is greater than the expected gain.
15.	Paliszkiewicz (2013)	Trust is the belief that another party: (1) will not act in a way that is harmful to the trusting firm, (2) will act in such a way that it is beneficial to the trusting firm, (3) will act reliably, and (4) will behave or respond in a predictable and mutually acceptable manner.

Source: Based on: Wrightsman, L. S., *Journal of Personality and Social Psychology*, 4, 328–332, 1966; Rotter, J. B., *Journal of Personality*, 35, 651–665, 1967; Gibb, J. R., *Trust: A new view of personal and organizational development*, Guild of Tutors Press, International College, Los Angeles, 1978; Sako, M., *Prices, quality and trust, inter-firm relations in Britain & Japan*, Cambridge University Press, Cambridge, 1992; Morgan, R. M., & Hunt, S. D., *Journal of Marketing*, 58, 20–38, 1994; Lewicki, R. J., & Bunker, B. B., Trust in relationships: A model of trust development and decline. In B. B. Bunker & J. Z. Rubin [eds.], *Conflict, cooperation and justice*, San Francisco, Jossey-Bass, 1995; Mayer, R. C. et al., *Academy of Management Review*, 20, 709–734, 1995; McAllister, D. J., *Academy of Management Journal*, 38, 24–59, 1995; Bidault, F., & Jarillo, J. C. Trust in economic transactions. In F. Bidault et al., [eds.], *Trust, firm and society*, London, Macmillan, 1997; Doney, P. M., & Cannon, J. P., *Journal of Marketing*, 61, 35–51, 1997; Gambetta, D., Can we trust trust? In D. Gambetta [ed.], *Trust: Making and breaking cooperative relations*, Oxfor, Basil Blackwell, 1998; Bhattacharya, R. et al., *Academy of Management Review*, 23, 459–472, 1998; Rousseau, D. M. et al., *Academy of Management Review*, 23, 393–404, 1998; Sztompka, P. *Trust: A sociological theory*, Cambridge University Press, Cambridge, 1999; Tatham, P., & Kovacs, G., *International Journal of Production Economics*, 126, 35–45, 2010; Paliszkiewicz, J. *Zaufanie w zarządzaniu* [Trust in management], Wydawnictwo Naukowe PWN, Warszawa, 2013.

Table 5.2 Types of Trust

S.N.	Authors	Types of Trust
1.	Zucker (1986)	Institutional-based, characteristic-based, process-based
2.	Rousseau et al. (1988)	Calculus-based, relational, institution-based
3.	Lewicki & Bunker (1996)	Calculus-based, knowledge-based, identification-based
4.	Jones & George (1998)	Conditional, unconditional
5.	Solomon & Flores (2001)	Basic, simple, blind, and authentic
6.	Bibb & Kourdi (2004)	Self-trust, rational, structural, transactional
7.	Paul & McDaniel (2004)	Calculative, competence, relational, integrated
8.	Siegrist, Gutscher, & Earle (2005); Ding, Veeman, & Adamowicz (2011)	General, specific
9.	Puusa & Tolvanen (2006)	Individual, group, systems
10.	Delhey, Newton, & Welzel (2011)	In-group, out-group
11.	Krot & Lewicka (2012)	Horizontal, vertical, institutional

Source: Based on: Zucker, L. G., *Research in Organizational Behavior*, 8, 53–111, 1986; Rousseau, D. M. et al., *Academy of Management Review*, 23, 393–404, 1998; Lewicki, R. J., & Bunker, B. B., Developing and maintaining trust in work relationships. In R. M. Kramer, & T. R. Tyler [eds.], *Trust in organizations, frontiers of theory and research*, Thousand Oaks, 1996, Sage; Jones, G., & George, J., *Academy of Management Review*, 23, 531–548, 1998; Solomon, R. C., & Flores, F. *Building trust in business, politics, relationships, and life*, Oxford University Press, Oxford, 2001; Bibb, S., & Kourdi, J., *Trust matters for organizational and personal success*, Palgrave Macmillan, Basingstoke, 2004; Paul, D. L., & McDaniel, R. R., Jr., *MIS Quarterly*, 28, 183–227, 2004; Siegrist, M. et al., *Journal of Risk Research*, 8, 145–156, 2005; Ding, Y. L. et al., *Agribusiness*, 27, 1–13, 2011; Puusa, A., & Tolvanen, U., *Electronic Journal of Business Ethics and Organization Studies*, 11, 29–31, 2006; Delhey, J. et al., *American Sociological Review*, 76, 786–807, 2011; Krot, K., & Lewicka, D., *International Journal of Electronic Business Management*, 10, 224–233, 2012.

Institutional-based trust, characteristic-based trust, and process-based trust were introduced by Zucker (1986). According to Zucker, institutional-based trust is related to formal societal structures depending on the firm—specific or individual attributes and on intermediary attributes. Characteristic-based trust is related to a person. Process-based trust is related to expected or past experience, for example, reputation.

Rousseau et al. (1988) described three basic types of trust: calculus-based trust, relational trust, and institution-based trust. Calculus-based trust attempts will calculate whether the relationship will pay off. Relational trust increases over time through repeated interactions that enable to create an experience and to predict the behavior of others in the future. Institution-based trust is related to institutions, cultural, and societal norms and with the trustor's feeling of security related to guarantees, regulations, or the legal system.

Lewicki and Bunker (1996) distinguish three levels of trust. The first level is calculus-based like in the proposition of Rousseau et al. (1988). The second level is called knowledge-based in which trust begins to develop based on previous experience. The third level of trust is identification-based. At this level, people know each other and may predict others' behaviors.

Conditional and unconditional types of trust are described by Jones and George (1998). Conditional trust is based on positive expectations on activities and intention. Unconditional trust occurs when individuals abandon the pretense of suspending belief. In this situation, this means that the relationship is significant.

Solomon and Flores (2001) classified trust as basic, simple, blind, and authentic. Basic trust is related to the disposition to trust. It can be described as the ability to create relationships with others. In simple trust, people do not have suspicions. They do not question each other's trustworthiness. Blind trust means that people think that other parties cannot act against them. Authentic trust is the most mature type of trust based on previous individual experiences.

According to Bibb and Kourdi (2004), there are four different types of trust: self-trust, rational trust, structural trust, and transactional trust. Self-trust is fundamental for building trust with other people. If people do not trust themselves, it is unlikely that others will trust them. Rational trust is based on previous experiences. Structural trust is the type of trust that people put in entire institutions. Transactional trust is specific and appears in a particular context.

Paul and McDaniel (2004) presented four types of trust: calculative, competence, relational, and integrated. Calculative trust is based on the economic exchange theory. Competence trust is based on expectations that another party is capable of doing what it says (Mayer et al., 1995; Mishra, 1996). Relational trust is the extent that one party feels a personal attachment to the other party and wants to behave properly (Jarvenpaa, Knoll, & Leidner, 1998; Mayer et al., 1995). An integrated type of trust is a combination of different types of trust, which are related to each other.

Siegrist et al. (2005) and Ding et al. (2011) described two types of trust. They are general trust and specific trust. General trust is presented as trust toward people in general and specific trust refers to trust specifically related to the given referent (for example, trust to the company). General trust is also referred to as basic trust (Brenkert, 1998; Solomon & Flores, 2001).

Puusa and Tolvanen (2006) classified trust in three levels: individual, group, and systems. At the first level, trust is related to interpersonal interactions; at the second level, trust is linked with project teams; and at the third level, trust relates to the reputation of the organization.

Delhey et al. (2011) distinguished two types of trust: in-group and out-group. In-group trust involves a narrow circle of familiar people, such as family, neighbors, and individuals known personally (particularistic trust). Out-group trust concerns with a wider circle of unfamiliar people (for example, people from different countries).

Krot and Lewicka (2012) presented horizontal trust (between co-workers), vertical trust (between managers and employees or between employees and managers), and institutional trust (between employees and organizations).

The Process of Building Trust in Organizations

Trust building can be defined as an interactive process in which people learn how to maintain trustworthiness in organizations. Trust is built up gradually and reinforced by previous positive experiences (Lewicki & Bunker, 1996). It needs time to be developed.

According to Bracey (2002), the leader may start building interpersonal trust using the following five TRUST principles: be Transparent, be Responsive, Use care, be Sincere, be Trustworthy. Galford and Drapeau (2002) presented the SEEKER model with the critical elements of trust building: Show that you understand the needs of the person and/or group; Establish the guiding principles of how you will operate; Explain the resources you will use in this work; Keep to the principles you have elaborated; Engage in constant, honest, two-way communication; Reinforce through consistent behaviors. According to Mishra and Mishra (2008), the main characteristics that help in trust building are reliability, openness, competence, and compassion. Sprenger (2004) listed reliability, consistency, predictability, keeping promises, fairness, loyalty, honesty, discretion, and credibility as the important traits in this area. According to Armour (2007), the characteristics include humility, integrity, truth, responsiveness, unblemished fair play, support and encouragement, and team care.

According to Boutros and Joseph (2007), trust should be carefully built, carefully maintained, and steadfastly renovated as needed. The foundation of trust is self-trust defined as the ability to trust oneself wisely and authentically (Lehrer, 1997). By using intuition, people can develop trust in themselves. Intuition has a significant role in trust building in business and management (Burke & Miller, 1999; Dane & Pratt, 2007; Hayes, Allinson, & Armstrong, 2004; Miles & Sadler-Smith, 2014;

Table 5.3 Actions to Build and Destroy Trust

Actions: To Build Trust	Actions: To Destroy Trust
Keeping promises Act fairly and consistently	Breaking promises
Honesty, openness	Manipulation of people
Kindness, loyalty	Unkindness, disloyalty
Apologies	Arrogance
Share important information about yourself	Do not be open to others
Forgiveness	Holding grudges
Engage in constant, honest, two-way communication	Lack of authentic communication
Reliability	Unreliability
Managing mutual expectations Support and encouragement Team care	Do not care about others
Show respect to others and create transparency Treat others as you would wish to be treated	Do not show respect to others

Source: Own elaboration, based on: Galford, R., & Drapeau, A. S., *The trusted leader. Bringing out the best in your people and your company,* Free Press, New York, 2002; Sprenger, R. K. *Trust. The best way to manage,* Cyan/Campus, London, 2004; Six, F., *The trouble with trust. The dynamics of interpersonal trust building,* MPG Books, Bodmin, 2005; Covey, S. M. R. (2009). Building trust. How the best leaders do it. *Leadership Excellence,* January; Armour, M. *Leadership and the power of trust. Creating a high-trust peak-performance organization,* Life Themes Press, Dallas, 2007; Boutros, A., & Joseph, C. B., *Physician Executive,* 33, 38–41, 2007; Mishra, A., & Mishra, K., *Trust is everything. Become the leader others will follow,* Aneil Mishra and Karen Mishra, Winston-Salem, 2008.

Miller & Ireland, 2005; Parikh, Neubauer, & Lank, 1994; Sadler-Smith & Shefy, 2004; Salas, Rosen, & Diaz Granados, 2010). Some actions that support and destroy trust in organizations are presented in Table 5.3.

The process of trust building is always related to a certain amount of risk. According to Six (2007), there are three main factors that make it difficult to build trust: the first is the interactive process that involves (at least) two individuals learning about each other's trustworthiness; the second is to build up trust gradually and incrementally, reinforced by previous trusting behavior and previous positive experiences; and the third is that there is no absolute certainty that trust will be honored. Recovering trust requires three separate actions (Boutros & Joseph, 2007): sincere apologies, permitting the affected person to influence you, and fulfilling the promise. Distrust affects different areas of the organization, for example, lack of communications, unclear policy of the organization, pressure on the workforce, and lack of ethics (Shockley-Zalabak, Morreale, & Hackman, 2010). Distrust usually appears in organizations where the future is insecure, workers are not treated fairly, and working conditions are inadequate (Kiefer, 2005; Wong, Wong, Ngo, & Lui, 2005). The costs of distrust are related to broken reputations, negative attitudes toward work, lower organizational performance, and making bad decisions.

Conclusion and Propositions for Future Research

The process of building and maintaining trust is very important in organizations. In the literature, we can find a lot of research, which confirms this importance. Trust affects many areas of organizational life in a positive way, for example, it stimulates collaboration (Anderson & Narus, 1990; Ganesan, 1994; Jap, 1999; Koohang, Paliszkiewicz, & Gołuchowski, 2017; Morgan & Hunt, 1994; Paliszkiewicz, 2013), it enables creating relationships (Ganesan, 1994; Morgan & Hunt, 1994; Ring & Van de Ven, 1992), it facilitates management coordination among different organizational units (Doney, Cannon, & Mullen, 1998; McAllister, 1995), and it improves the organizational climate (Das & Teng, 1998). The benefits of trust are obvious, but methods of building trust and maintaining it still remain a challenge to most people in organizations.

The theory described in this chapter requires empirical testing. There is a need for research related to how trust is built and re-built. Future research should focus on the costs and benefits of building and maintaining trust using people of different cultures to measure the level of trust and describe the meaning and importance of trust. The methods of measuring trust should be studied. A further challenge to the researcher is to study how trust influences intuition and decision-making.

References

Anderson, J. C., & Narus, J. A. (1990). A model of distributor firm and manufacturer firm working partnerships. *Journal of Marketing, 54*(1), 42–58.

Armour, M. (2007). *Leadership and the power of trust. Creating a high-trust peak-performance organization.* Dallas, TX: Life Themes Press.

Bhattacharya, R., Devinney, T. M., & Pillutla, M. M. (1998). A formal model of trust based on outcomes. *Academy of Management Review, 23*(3), 459–472.

Bibb, S., & Kourdi, J. (2004). *Trust matters for organizational and personal success.* New York: Palgrave Macmillan.

Bidault, F., & Jarillo, J. C. (1997). Trust in economic transactions. In F. Bidault, P. Gomez, & G. Marion (Eds.), *Trust, firm and society* (pp. 81–94). London: Macmillan.

Boutros, A., & Joseph, C. B. (2007). Building, maintaining and recovering trust: A core leadership competency. *Physician Executive, 33*(1), 38–41.

Bracey, H. (2002). *Building trust. How to get it! How to keep it!* Taylorsville, GA: Hyler Bracey.

Brenkert, G. G. (1998). Trust, morality and international business. *Business Ethics Quarterly, 8*(2), 293–317.

Burke, L. A., & Miller, M. K. (1999). Taking the mystery out of intuitive decision making. *Academy of Management Executive, 13*(4), 91–99.

Costigan, R. D., Ilter, S. S., & Berman, J. J. (1998). A multi-dimensional study of trust in organizations. *Journal of Managerial Issues, 10*, 303–317.

Covey, S. M. R. (2009). Building trust. How the best leaders do it. *Leadership Excellence,* January.

Dane, E., & Pratt, M. (2007). Exploring intuition and its role in managerial decision making. *Academy of Management Review, 32*(1), 33–54.

Das, T. K., & Teng, B. S. (1998). Between trust and control: Developing confidence in partner cooperation in alliances. *Academy of Management Review, 23*(3), 491–512.

Delhey, J., Newton, K., & Welzel, C. (2011). How general is trust in "most people"? Solving the radius of trust problem and deriving a better measure. *American Sociological Review, 76*(5), 786–807.

Ding, Y. L., Veeman, M. M., & Adamowicz, W. L. (2011). The impact of generalized trust and trust in food system on choice of a functional GM food. *Agribusiness, 27*, 1–13.

Doney, P. M., & Cannon, J. P. (1997). An examination of the nature of trust in buyer-seller relationships. *Journal of Marketing, 61*(April), 35–51.

Doney, P. M., Cannon, J. P., & Mullen, M. R. (1998). Understanding the influence of national culture on the development of trust. *Academy of Management Review, 23*(3), 601–620.

Eikeland, T. B. (2015). Emergent trust and work life relationships: How to approach the relational moment of trust. *Nordic Journal of Working Life Studies, 5*(3), 59–77.

Galford, R., & Drapeau, A. S. (2002). *The trusted leader. Bringing out the best in your people and your company.* New York, NY: Free Press.

Gambetta, D. (1998). Can we trust trust? In D. Gambetta (Ed.), *Trust: Making and breaking cooperative relations* (pp. 213–238). Oxfor: Basil Blackwell.

Ganesan, S. (1994). Determinants of long-term orientation in buyer—Seller relationships. *Journal of Marketing, 58*(2), 1–19.

Gibb, J. R. (1978). *Trust, A new view of personal and organizational development.* Los Angeles, CA: Guild of Tutors Press, International College.

Hakanen, M., & Soudunsaari, A. (2012). Building trust in high-performing teams. *Technology Innovation Management Review, 2*(6), 38–41.

Hayes, J., Allinson, C. W., & Armstrong, S. J. (2004). Intuition, women managers and gendered stereotypes. *Personnel Review, 33*(4), 403–417.

Jap, S. D. (1999). Pie-expansion efforts: Collaboration processes in buyer—Supplier relationships. *Journal of Marketing Research, 36*(4), 461–475.

Jarvenpaa, S. L., Knoll, K., & Leidner, D. E. (1998). Is anybody out there? Antecedents of trust in global virtual teams. *Journal of Management Information Systems, 14*(4), 29–64.

Jones, G., & George, J. (1998). The experience and evolution of trust: Implications for cooperation and teamwork. *Academy of Management Review, 23*(3), 531–548.

Kiefer, T. (2005). Feeling bad: Antecedents and consequences of negative emotions in ongoing change. *Journal of Organizational Behavior, 26*, 875–897.

Koohang, A., Paliszkiewicz, J., & Gołuchowski, J. (2017). The impact of leadership on trust, knowledge management, and organizational performance: A research model. *Industrial Management & Data Systems, 117*(3), 521–537.

Krishnan, R., Martin, X., & Noorderhaven, N. G. (2006). When does trust matter to alliance performance? *Academy of Management Journal, 49*(5), 894–917.

Krot, K., & Lewicka, D. (2012). The importance of trust in manager-employee relationships. *International Journal of Electronic Business Management, 10*(3), 224–233.

Lehrer, K. (1997). *Self-trust: A study of reason, knowledge, and autonomy.* New York, NY: Oxford University Press.

Lewicka, D., & Krot, K. (2015). The model of HRM-trust-commitment relationships. *Industrial Management & Data Systems, 115*(8), 1457–1480.

Lewicki, R. J., & Bunker, B. B. (1995). Trust in relationships: A model of trust development and decline. In B. B. Bunker & J. Z. Rubin (Eds.), *Conflict, cooperation and justice* (pp. 131–145). San Francisco, CA: Jossey-Bass.

Lewicki, R. J., & Bunker, B. B. (1996). Developing and maintaining trust in work relationships. In R. M. Kramer, & T. R. Tyler (Eds.), *Trust in organizations, frontiers of theory and research* (pp. 114–139). Thousand Oaks, CA: Sage.

Lewis, J. D., & Weigert, A. (1985). Trust as a social reality. *Social Forces, 63*, 967–984.

Li, J. J., Poppo, L., & Zhou, K. Z. (2010). Relational mechanisms, formal contracts, and local knowledge acquisition by international subsidiaries. *Strategic Management Journal, 31*(4), 349–370.

Malhotra, D., & Murnighan, J. K. (2002). The effects of contracts on interpersonal trust. *Administrative Science Quarterly, 47*(3), 534–559.

Mayer, R. C., Davis, J. H., & Schoorman, F. D. (1995). An integrative model of organizational trust. *Academy of Management Review, 20*(3), 709–734.

McAllister, D. J. (1995). Affect and cognition based trust as foundations for interpersonal cooperation in organizations. *Academy of Management Journal, 38*(1), 24–59.

Miles, A., & Sadler-Smith, E. (2014). With recruitment I always feel I need to listen to my gut: The role of intuition in employee selection. *Personnel Review, 43*(4), 606–627.

Miller, C. C., & Ireland, R. D. (2005). Intuition in strategic decision making: Friend or foe in the fast-paced 21st century? *Academy of Management Executive, 19*(1), 19–30.

Mishra, A. K. (1996). Organizational responses to crisis: The centrality of trust. In R. Kramer & T. Tyler (Eds.), *Trust in organizations: Frontiers of theory and research* (pp. 261–287). Thousand Oaks, CA: Sage Publications.

Mishra, A., & Mishra, K. (2008). *Trust is everything. Become the leader others will follow.* Winston-Salem, NC: Aneil Mishra and Karen Mishra.

Morgan, R. M., & Hunt, S. D. (1994). The commitment-trust theory of relationship marketing. *Journal of Marketing, 58*(3), 20–38.

Muethel, M., & Bond, M. H. (2013). National context and individual employees' trust of the out-group: The role of societal trust. *Journal of International Business Studies, 44*(4), 312–333.

Nooteboom, B. (2002). *Trust: Forms, foundations, functions, failures and figures.* Cheltenham: Edward Elgar.

Paliszkiewicz, J. (2013). *Zaufanie w zarządzaniu* [Trust in management]. Warszawa: Wydawnictwo Naukowe PWN.

Paliszkiewicz, J., Gołuchowski, J., & Koohang, A. (2015). Leadership, trust, and knowledge management in relation to organizational performance: Developing an instrument. *Online Journal of Applied Knowledge Management, 3*(2), 19–35.

Paliszkiewicz, J., & Koohang, A. (2016). *Social media and trust: A multinational study of university students.* Santa Rosa, CA: Informing Science Press.

Paliszkiewicz, J., Koohang, A., & Horn Nord, J. (2014). Management trust, organizational trust, and organizational performance: Empirical validation of an instrument. *Online Journal of Applied Knowledge Management, 2*(1), 28–39.

Pangil, F., & Chan, J. M. (2014). The mediating effect of knowledge sharing on the relationship between trust and virtual team effectiveness. *Journal of Knowledge Management, 18*(1), 92–106.

Parikh, J., Neubauer, F., & Lank, A. G. (1994). *Intuition: The new frontier of management.* London: Blackwell.

Paul, D. L., & McDaniel, R. R., Jr. (2004). A field study of the effect of interpersonal trust on virtual collaborative relationship performance. *MIS Quarterly, 28*(2), 183–227.

Puusa, A., & Tolvanen, U. (2006). Organizational identity and trust. *Electronic Journal of Business Ethics and Organization Studies, 11*(2), 29–31.

Ring, P. S., & Van den Ven, A. H. (1992). Structuring cooperative relationships between organizations. *Strategic Management Journal, 13*(7), 483–498.

Rotter, J. B. (1967). A New scale for the measurement of interpersonal trust. *Journal of Personality, 35,* 651–665.

Rousseau, D. M., Sitkin, S. B., Burt, R. S., & Camerer, C. (1998). Not so different after all: Across-discipline view of trust. *Academy of Management Review, 23*(3), 393–404.

Sadler-Smith, E., & Shefy, E. (2004). The intuitive executive: Understanding and applying 'gut feel' in decision making. *Academy of Management Executive, 18*(4), 76–91.

Sako, M. (1992). *Prices, quality and trust, inter-firm relations in Britain & Japan.* Cambridge: Cambridge University Press.

Salas, E., Rosen, M., & Diaz Granados, D. (2010). Expertise-based intuition and decision making in organizations. *Journal of Management, 36*(4), 941–973.

Shockley-Zalabak, P. S., Morreale, S. P., & Hackman, M. Z. (2010). *Building the high-trust organization.* San Francisco, CA: Jossey-Bass.

Siegrist, M., Gutscher, H., & Earle, T. C. (2005). Perception of risk: The influence of general trust and general confidence. *Journal of Risk Research, 8*(2), 145–156.

Six, F. (2004). *Trust and trouble. Building interpersonal trust within organizations.* Rotterdam: Erasmus Research Institute of Management.

Six, F. (2005). *The trouble with trust. The dynamics of interpersonal trust building.* Bodmin: MPG Books.

Six, F. E. (2007). Building interpersonal trust within organizations: A relational signaling perspective. *Journal of Management & Governance, 11*(3), 285–309.

Solomon, R. C., & Flores, F. (2001). *Building trust in business, politics, relationships, and life.* Oxford: Oxford University Press.

Sprenger, R. K. (2004). *Trust. The best way to manage.* London: Cyan/Campus.

Sztompka, P. (1999). *Trust: A sociological theory.* Cambridge: Cambridge University Press.

Taddeo, M. (2009). Defining trust and E-trust: From old theories to new problems. *International Journal of Technology and Human Interaction, 5*(2), 23–35.

Tatham, P., & Kovacs, G. (2010). The application of "swift trust" to humanitarian logistics. *International Journal of Production Economics, 126*(1), 35–45.

Tyler, T. R. (2003). Trust within organisations. *Personnel Review, 32*(5), 556–568.

Wong, Y. T., Wong, C. S., Ngo, H. Y., & Lui, H. K. (2005). Different responses to job insecurity of Chinese workers in joint ventures and state-owned enterprises. *Human Relations, 58,* 1391–1418.

Wrightsman, L. S. (1966). Personality and attitudinal correlates of trusting and trustworthy behaviors in a two-person game. *Journal of Personality and Social Psychology, 4,* 328–332.

Yoon, A. (2014). End users' trust in data repositories: Definition and influences on trust development. *Archival Science, 14*(1), 17–34.

Young, L., & Daniel, K. (2003). Affectual trust in the workplace. *International Journal of Human Resource Management, 14*(1), 139–155.

Zucker, L. G. (1986). Production of trust: Institutional sources of economic structure, 1840–1920. *Research in Organizational Behavior, 8,* 53–111.

Chapter 6

Trust, Knowledge Management, and Organizational Performance: Predictors of Success in Leadership

Alex Koohang, Joanna Paliszkiewicz,
and Jerzy Gołuchowski

Contents

Introduction

Leadership is a multifaceted topic in today's globalized world (Gandolfi & Stone, 2016). It has a fundamental role in directing and shaping organizations by providing a sense of direction, vision, and purpose for all members. New questions and challenges in management continue to emerge with regard to effective leadership that can influence the organizational culture, knowledge sharing, employees' positive attitudes toward their jobs, and create ethical norms (e.g., Barling, Weber, & Kelloway, 1996; Charbonneau, Barling, & Kelloway, 2001; Grojean, Resick, Dickson, & Smith, 2004; Howell & Avolio, 1993; Mayer, Kuenzi, Greenbaum, Bardes, & Salvador, 2009; Schaubroeck et al., 2012).

Leadership is often described as a process of influence toward the accomplishment of objectives of the organization (Bass, 1960; Katz & Kahn, 1966). This view of leadership generally focuses on the relationship between a leader and follower, but not on what conditions need to be in place for successful leadership.

Koohang, Paliszkiewicz, and Gołuchowski (2017) proposed a research model that explored the influence of effective leadership on organizational trust among employees, knowledge management processes within organizations, and organizational performance. The results of their study showed that effective leadership (i.e., leading organization, leading people, and leading self), increased trust among employees, advanced the successful implementation of knowledge management processes, and in turn improved organizational performance.

The goal of this chapter is to examine whether effective leadership (leading organization, leading people, and leading self) as a predicting variable influences the three predictor variables of trust, knowledge management, and organizational performance within organizations.

Review of the Literature

Trust

Trust is vital in all spheres of human and professional life. Trust has been a topic of research in many areas of science, for example, in psychology, sociology, and management. Thus, in the literature, there are different propositions of the definition of trust.

Psychologists define it as a personal trait (Erikson, 1968; Rotter, 1967) or as a feeling, which is very close to love (Gibb, 1978). People vary in terms of how much and when they are willing to trust (Das & Teng, 2004). Differentiation in the propensity to trust among people can be related to their developmental experiences, personality types, and cultural backgrounds (Mayer, Davis, & Schoorman, 1995), but it also depends on previous personal experiences and socialization.

Sociologists describe trust as a part of a social structure (e.g., Garfinkel, 1967; Lewis & Weigert, 1985a; Shapiro, 1987). Some, such as Lewis and Weigert (1985a) insist that trust cannot be considered only as a personality trait because trust is a complex and multidimensional phenomenon. Sociologists research trust as a fundament of friendship, families, or nations. According to Sztompka (1999), trust is the expectation that other people, groups, or institutions with whom we interact will act in ways conducive to our well-being.

Trust is important in professional life (Grudzewski, Hejduk, Sankowska, & Wańtuchowicz, 2007). In the context of management, Mayer et al. (1995, p. 712) characterize trust as "the willingness of a party to be vulnerable to the actions of another party based on the expectation that the other will perform a particular action important to the trustor, irrespective of the [trustor's] ability to monitor or control that other party." Paliszkiewicz (2013) defines trust as the belief and optimistic expectation that another party will act in such a way that it is beneficial to the trusting party, and will act reliably and will behave or respond in a predictable and mutually acceptable manner. Morgan and Hunt (1994, p. 23) define trust as "one party's confidence in an exchange partner's reliability and integrity."

The importance of trust in management has been described in literature. There are many aspects in organizations on which trust influences, for example: increase team performance (Costa, 2003; Costa, Roe, & Taillieu, 2001; Klimoski & Karol, 1976); limit transaction costs (Doney et al., 1998; Dore, 1983; Handy, 2005); stimulate collaboration among partners (Anderson & Narus, 1990; Ganesan, 1994; Jap, 1999; Morgan & Hunt, 1994; Paliszkiewicz, 2013); facilitate long-term relationships (Ganesan, 1994; Morgan & Hunt, 1994; Ring & Van de Ven, 1992); facilitate management coordination among different organizational units (Doney et al., 1998; McAllister, 1995); contribute to effective implementation of strategies (Doney et al., 1998); facilitate more open communication and sharing of knowledge (Chowdhury, 2005; Lewis & Weigert, 1985b; McAllister, 1995; Seppanen, Blomqvist, & Sundqvist, 2007); and enhance organizational performance (Paliszkiewicz, Koohang, & Horn Nord, 2014; Seppanen et al., 2007).

Knowledge Management

Knowledge management has been a popular research topic for many scholars (Davenport & Prusak, 1998; Liebowitz, 2008, 2012, 2016; Nonaka & Takeuchi, 1995; Sveiby, 1987). The focus of knowledge management is on how an organization's knowledge resources are used to contribute to the success of an organization (Holsapple, Hsiao, & Oh, 2016) and provide a sustainable competitive advantage (e.g., Davenport & Prusak, 1998; Foss & Pedersen, 2002; Grant, 1996; Spender & Grant,

1996; Wang & Noe, 2010). According to a number of researchers, knowledge is a critical resource that should be managed strategically (Grant, 1996; Spender, 1996; Teece, 1998).

According to Brooking (1999, p. 154) the function of knowledge management is ". . . to guard and grow knowledge owned by individuals, and where possible, transfer the asset into a form where it can be more readily shared by other employees in the company." Also, McInerney (2002) in his definition underlines an effort of knowledge management to increase useful knowledge within the organization that results in encouragement to communicate, offering opportunities to learn, and promoting the sharing of appropriate knowledge artifacts.

Holsapple and Joshi (2004, p. 596) define knowledge management as ". . . an entity's systematic and deliberate efforts to expand, cultivate and apply available knowledge in ways that add value to the entity, in the sense of positive results in accomplishing its objectives or fulfilling its purpose."

In the literature, there are many definitions of knowledge management. According to Awad and Ghaziri (2004), each definition of knowledge management contains several integral parts: using accessible knowledge from outside sources; embedding and storing knowledge in business processes, products, and services; representing knowledge in databases and documents; promoting knowledge growth through the organization's culture and incentives; transferring and sharing knowledge throughout the organization; assessing the value of knowledge assets; and impacting knowledge on a regular basis.

Several studies have addressed knowledge management processes. They divide knowledge management into several processes, for example:

- Capturing, distributing, and using knowledge (Davenport, 1994)
- Creation, capture, and use of knowledge (Bassie, 1997)
- Creation, extraction, transformation, and storage of knowledge (Horwitch & Armacost, 2002)
- Acquisition, dissemination, and responsiveness (Darroch, 2003)
- Creating, gathering, organizing, storing, diffusing, using, and exploiting knowledge (Chong & Choi, 2005)
- Creation, accumulation, sharing, utilization, and internalization of knowledge (Lee, Lee, & Kang, 2005)

Paliszkiewicz (2007) has proposed that knowledge management includes five processes. They are localization (all activities that indicate where knowledge exists); usage of knowledge (creating a set of roles and skills to effectively use knowledge); knowledge acquisition and development (the culture of embracing the knowledge that is acquired and developed); knowledge codification (the ability to successfully and continuously re-use the knowledge that organizations capture); and knowledge transfer (transmission of knowledge and use of transmitted knowledge).

Managers set the tone for how employees should perform. Within the framework of knowledge management practices, managers are expected to create a culture or an organizational climate that sets work norms and values that enable knowledge management and transfer of knowledge to produce value-added products and services.

Organizational Performance

Organizational performance is concerned with measures of how well a task is completed relative to criteria established for organizational effectiveness. It measures the progress and development and how well an organization is accomplishing its goals and objectives.

Different indicators of organizational performance are documented in the literature, that is, financial (Parmenter, 2009), quality (De Toni & Tonchia, 2001; Gosselin, 2005), reliability (White, 1996), productivity (Sinclair & Zairi, 1995), customer satisfaction (Neely, Gregory, & Platts, 2005), time (White, 1996), learning and growth (Sadler-Smith, Spicer, & Chaston, 2001), and leadership (Lieberson and O'Connor, 1972; Thomas, 1988). The most widely used organizational performance indicators in the literature are those outlined by Sink and Tuttle (1989). These indicators include effectiveness, efficiency, quality, productivity, quality of work life, innovation, and profitability/budget ability.

Leadership

The topic or leadership has interested writers from centuries ranging from the early Greek philosophers such as Plato and Socrates to today's researchers (Gandolfi & Stone, 2016). Leadership relates to how a leader chooses to lead and how his or her behavior influences an organization.

Leadership can be defined as "the process of interactive influence that occurs when, in a given context, some people accept someone as their leader to achieve common goals" (Silva, 2016, p. 3).

Winston and Patterson (2006, p. 7) describe leaders as ". . . one or more people who selects, equips, trains, and influences one or more follower(s) who have diverse gifts, abilities, and skills and focuses the follower(s) to the organization's mission and objectives causing the follower(s) to willingly and enthusiastically expand spiritual, emotional, and physical energy in a concerted coordinated effort to achieve the organizational mission and objectives."

Kouzes and Posner (2007) present five key attributes for effective leadership: (1) model the way, (2) inspire a shared vision, (3) challenge the process, (4) enable others to act, and (5) encourage the heart.

According to Debowski (2006), the characteristics of good leaders tend to reflect four key themes: (1) the capacity to explain and clarify the organization's purpose and priorities, (2) the development of the culture within which workers operate, (3) the creation and maintenance of good people practices to facilitate effective work, and (4) the encouragement of high performance in the work setting.

Effective leadership within an organization is often viewed as the foundation for organizational performance and growth. In organizations that don't have strong leaders, it is more difficult to meet performance expectations.

Purpose

Leadership effectiveness is vital to the success of any organization. Effective leadership contributes to employees' positive attitudes; employees' positive work climate; employees' willingness to share information; and the employees' accomplishment of positive team performance (cf. Paliszkiewicz, Gołuchowski, & Koohang, 2015). Koohang et al. (2017) proposed a research model that explored leadership effectiveness influence on organizational trust, knowledge management processes within organizations, and organizational performance. The findings of their study implied that effective leadership (i.e., leading organization, leading people, and leading self) contributed to increased trust among employees, advanced the successful implementation of knowledge management processes, and in sequence improved organizational performance. In the present study, we attempt to examine whether effective leadership as a predicting variable influences the three predictor variables of trust, knowledge management, and organizational performance within organizations. Both the predicting and the predictor variables are defined by certain characteristics, that is, effective leadership has fifteen characteristics grouped into three constructs. Trust includes ten characteristics. Knowledge management has five characteristics and organizational performance includes seven characteristics.

Fifteen leadership characteristics were outlined in a study by Paliszkiewicz et al. (2015). Koohang et al. (2017, p. 526) grouped these characteristics into three leadership constructs: leading organization, leading people, and leading self. The leadership: leading organization encompasses the characteristics that assure organizational advancement. The definitions of these characteristics are presented below.

Leadership: Leading Organization

1. *Change*: Change is required, inevitable and continuous in any organizations. Organizational change requires sound leadership.
2. *Innovation*: Innovation moves an organization forward. Leading innovation is the ability of a leader to lead innovation within an organization.
3. *Influence*: A leader has the ability to influence and be flexible.
4. *Diversity and Inclusion*: A leader has the ability to value individual differences. A leader respects and appreciates diversity and inclusion (Koohang et al., 2017).

The leadership: Leading people consists of the characteristics that enhance productivity among employees. The definitions of these characteristics are presented.

Leadership: Leading People

1. *Motivation*: Motivating employees brings about productivity. A leader's task is to motivate and bring out the best in employees.
2. *Listening*: A leader has the ability to be a good listener. A good listener will put employees at ease and make them comfortable.
3. *Empowerment*: A leader has the ability to empower others to do their jobs. Empowering creates autonomy and responsibility; therefore, employees can participate in decision-making within organizations.
4. *Interpersonal Communication*: A leader has the ability to communicate effectively. He or she is interpersonal savvy.
5. *Building Relationships*: A leader has the ability to build and sustain relationships among employees.
6. *Conflict*: A leader is not afraid of conflicts. A leader's attitude toward conflict must be positive. A leader does not avoid conflict and considers conflict as an opportunity to improve a situation (Koohang et al., 2017).

The leadership: Leading self holds the characteristics that augment the development of a leader. The definitions of these characteristics are presented below.

Leadership: Leading Self

1. *Values/Principles*: A leader must be grounded in values and principles. He or she makes decisions and solve problems based on his/her values and principles.
2. *Self-Awareness*: A leader has the ability to be self-aware. A leader must be conscious and mindful of everyone within an organization.
3. *Feedback*: A leader is comfortable to seek feedback from employees. A leader uses the feedback for self-improvement.
4. *Managing Time*: A leader has the ability to effectively manage time.
5. *Learning*: A leader continuously seeks the opportunity to learn. A leader seeks new knowledge, modify existing knowledge, and apply what he/she learns to situations for improvement (Koohang et al., 2017).

Ten characteristics of trust were selected from Paliszkiewicz et al. (2015) and Koohang et al. (2017). These characteristics are ability/competence, benevolence, communication, congruency, consistency, dependability, integrity, openness, reliability, and transparency. The definitions of these characteristics are as follows:

1. *Ability/Competence*: One's demonstration of ability and competence (knowledge, skills, aptitude, qualification) lead to improved trust.
2. *Benevolence*: One's expression of compassion and empathy.
3. *Communication*: One's ability to effectively and constantly communicate (verbal, nonverbal, written, and visual).

4. *Congruency*: Displaying the attitude and enthusiasm of partnership and association among people.
5. *Consistency*: One's demonstration of consistency in performing various tasks.
6. *Dependability*: Exhibiting dedication, truthfulness, responsibility, and trustworthiness.
7. *Integrity*: Displaying honesty and exhibiting moral and ethical principles.
8. *Openness*: Showing acceptance and broad-mindedness.
9. *Reliability*: Exhibiting the ability to be depended on in performing tasks.
10. *Transparency*: One's ability to be transparent—not to hide or block information that is needed to perform tasks (cf. Paliszkiewicz et al., 2015).

The five knowledge management characteristics were chosen from Paliszkiewicz (2007). These characteristics include localization, usage of knowledge, knowledge acquisition and development, knowledge codification, and knowledge transfer. The definitions of these characteristics are presented as follows:

1. *Localization*: Includes all activities that indicate where knowledge exists.
2. *Usage of Knowledge*: Creating a set of roles and skills to effectively use knowledge.
3. *Knowledge Acquisition and Development*: The culture of embracing the knowledge that is acquired and developed.
4. *Knowledge Codification*: The ability to successfully and continuously re-use the knowledge that organizations capture.
5. *Knowledge Transfer*: Transmission of knowledge and use of the transmitted knowledge.

The seven organizational performance characteristics were chosen from Sink and Tuttle (1989). These characteristics include effectiveness, efficiency, productivity, quality, quality of work life, innovation, and profitability. The definitions of these characteristics are presented as follows:

1. *Effectiveness*: An output measure—the ratio of the expected output to the actual output
2. *Efficiency*: An input measure—the ratio of the expected input to the actual input.
3. *Quality*: Quality is the key to success of every organization. The quality is checked mainly at three levels input, output, and throughput or process quality. It can include actual input/output versus the expected accuracy, timeliness, and so on.
4. *Productivity*: Ratio of output to input.
5. *Quality of Work Life*: Employee attitudes to work.
6. *Innovation*: Measures the organization's success in creating change
7. *Profitability/Budget Ability*: An outcome to input ratio (Sink & Tuttle, 1989).

As stated earlier, we attempt to find out which of the predictor variables (trust, knowledge management, and organizational performance) are most influential in predicting each of the leadership variables (leading organization, leading people, and leading self). Therefore, the following research questions (RQ) are stated:

RQ1: Which of the three predictor variables (*trust, knowledge management, and organizational performance*) are most influential in predicting leadership: leading organization?

RQ2: Which of the three predictor variables (*trust, knowledge management, and organizational performance*) are most influential in predicting leadership: leading people?

RQ3: Which of the three predictor variables (*trust, knowledge management, and organizational performance*) are most influential in predicting leadership: leading self?

Methodology

The Instrument

The survey instrument was originally developed by Paliszkiewicz et al. (2015) with four constructs—leadership, trust, knowledge management, and organizational performance. Koohang et al. (2017) refined the leadership construct of the instrument to include three separate constructs: leading organization, leading people, and leading self. The refined instrument, therefore, includes six constructs: (1) leadership: leading organization, (2) leadership: leading people, (3) leadership: leading self, (4) trust, (5) knowledge management, and (6) organizational performance. The items for each construct are shown in Table 6.1.

Participants

The instrument was administered face-to-face to 116 subjects from various companies in Poland. The subjects were from all three levels of management: Senior/Top Level Management (52.6%), Middle Level Management (32.8%), and Supervisory/Lower Level Management (14.7%). They were male (77%) and female (23%), working in both private (26%) and public (74%) companies. All subjects were over the age of 25. The subjects were assured confidentiality and anonymity.

Data Analysis

For each research question, we used multiple regression analysis (the Enter method, where all independent variables enter the model one at a time) to determine which of the independent variables can best predict the dependent variable.

Table 6.1   Items of the Instrument

Leadership: Leading Organization Construct

1. *Change*: A leader must lead change within an organization.
2. *Innovation*: It is necessary for a leader to lead innovation within an organization.
3. *Influence*: The ability of a leader to positively shape the organization, people, and self by setting a vision, translating it into realistic business strategies, and expecting outcomes.
4. *Diversity and Inclusion*: A leader values and respects diversity and inclusion within an organization. Diversity brings about innovation. Inclusion ensures the right conditions for all, working together to enhance organizational effectiveness.

Leadership: Leading People Construct

1. *Motivation*: A leader must motivate and bring out the best in people.
2. *Listening*: A leader must empower others to do their jobs.
3. *Empowerment*: A leader must be a good listener and put people at ease.
4. *Interpersonal Communication*: A leader's interpersonal communication is necessary to bring people together to work effectively.
5. *Building Relationships*: A leader must build and maintain relationships with subordinates.
6. *Conflict*: A leader should not be afraid of conflict—a leader's attitude should be that conflict is "good" and should not be avoided.

Leadership: Leading Self Construct

1. *Values/Principles*: In making decisions, a leader must be grounded in values and principles.
2. *Self-Awareness*: A leader must be self-aware (knows his or her strengths and weaknesses and is willing to improve).
3. *Feedback*: A leader must seek and use feedback from others
4. *Managing Time*: A leader must know how to manage time efficiently.
5. *Learning*: A leader must seek the opportunity to learn continuously.

Trust Construct

1. *Competence*: A leader's ability and competence lead to improved trust among people.
2. *Benevolence*: Compassion and empathy demonstrated by a leader build trust among people.
3. *Communication*: A leader's sound and constant communication (verbal, nonverbal, written, and visual) improve trust among people.
4. *Congruency*: The attitude of partnership and association demonstrated by a leader build trust among people.
5. *Consistency*: Consistency in doing things by a leader brings about trust among people.
6. *Dependability*: Exhibiting dependability by a leader develops and creates trust among people.

Table 6.1   Items of the Instrument (*Continued*)

7. *Integrity*: A leader's honesty and principle contribute to elevated trust among people.
8. *Openness*: Acceptance and broad-mindedness demonstrated by a leader contribute to increased trust among people.
9. *Reliability*: Exhibiting reliability by a leader develops and creates trust among people.
10. *Transparency*: A leader's transparency is central to building trust among people.

Knowledge Management Construct

1. *Localization*: In any organization, knowledge must be localized to include all activities that indicate where knowledge exists.
2. *Usage of Knowledge*: Successful usage of knowledge depends on creating a set of roles and skills in organizations that encourage efficient use of knowledge.
3. *Knowledge Acquisition and Development*: The culture of embracing the knowledge that is acquired and developed is important in gaining the competitive advantage.
4. *Knowledge Codification*: Organizations must be able to successfully and continuously re-use the knowledge they capture.
5. *Knowledge Transfer*: Transmission of knowledge and use of the transmitted knowledge in any organization is vital to gaining the competitive advantage.

Organizational Performance Construct

1. *Effectiveness*: The ability to produce the desired result should be an important part of any organization.
2. *Efficiency*: The ability to accomplish a job/task with a minimum expenditure of time and effort should be central to any organizations
3. *Quality*: The quality of a product (as a measure of excellence and state of being free from defects, deficiencies, and significant variations) brings about the competitive advantage to any organization.
4. *Productivity*: The ability to resourcefully generate, create, enhance, and/or produce goods and services is vital.
5. *The Quality of Work Life*: The opportunity that is given to employees to improve their personal lives through their work environment and experiences can contribute to an organization's competitive advantage.
6. *Innovation*: The process of transforming an idea/invention into a product or service that creates value is vital to an organization's survival.
7. *Profitability*: A financial profit or gain gives an organization the ability to do more to gain the competitive advantage.

Source: Koohang et al., *Industrial Management and Data Systems*, 2017.

Note: The measuring scale was as follows: 7 = Completely agree; 6 = Mostly agree; 5 = Somewhat agree; 4 = Neither agree nor disagree; 3 = Somewhat disagree; 2 = Mostly disagree; 1 = Completely disagree.

The independent variables (predictor variables) were IV_1 = trust, IV_2 = knowledge management, and IV_3 = organizational performance (for all three RQs). The dependent variables (predicting variables) were DV_1 = leadership: leading organization (for RQ1), DV_2 = leadership: leading people (for RQ2), and DV_3 = leadership: leading self (for RQ3).

The multiple regression analysis produces a coefficient table to interpret the results for each research question. Prior to interpreting the results in the coefficient table, the test of multicollinearity is performed to examine the tolerance level and the variance inflation factor (VIF). The presence of multicollinearity constraints the size of R shows overlapping information and escalates the regression coefficients, thus prediction can be unstable (Stevens, 2001). To confirm the nonexistence of multicollinearity among the IVs, the tolerance level values for all IVs should be above .1, and the (VIF) values for all IVs should not be greater than 10. Once the nonexistence of multicollinearity is established, we proceed to interpret (1) the model summary, (2) the ANOVA, and (3) the coefficients table.

The model summary includes multiple correlations (R), squared multiple correlations (R^2), and adjusted squared multiple correlations (R^2_{adj}). The model summary point to how well the independent variables predict the dependent variable (Stevens, 2001).

The ANOVA table shows the F test and the corresponding level of significance. If the F test is significant, then the relationship between the dependent variable and the independent variables is linear, therefore, the model significantly predicts the dependent variable (Stevens, 2001).

The coefficients table includes the beta weights, t and p values for the independent variables, which are examined to determine the predictor variables that are most influential in predicting the dependent variable (Stevens, 2001).

Results

RQ1: Which of the three predictor variables (trust, knowledge management, and organizational performance) are most influential in predicting leadership: leading organization?

The test of multicollinearity confirmed the nonexistence of multicollinearity among the IVs. The tolerance level values for all IVs were above .1, and the VIF values for all IVs were greater than 10. The values were as follows: IV_1 = trust (TL = .474, VIF = 2.108), IV_2 = knowledge management (TL = .555, VIF =1.802), and IV_3 = organizational performance (TL = .519, VIF =1.926). Once the nonexistence of multicollinearity was established, we then interpreted (1) the model summary, (2) the ANOVA, and (3) the coefficients table.

The model summary revealed the results of multiple correlation (R = .759), squared multiple correlation (R^2 = .575), and the adjusted squared multiple correlation (R^2_{adj} = .564). These results indicated that the three independent variables of trust, knowledge management, and organizational performance can soundly predict the dependent variable of leadership: leading organization.

Table 6.2   Coefficients

	Unstandardized Coefficients		Standardized Coefficients		
	B	Std. Error	Beta	t	Sig.
(Constant)	−.249	.508		−.491	.624
Trust	.391	.106	.330	3.692	**.000**
Knowledge management	.447	.096	.384	4.643	**.000**
Organizational performance	.178	.099	.154	1.805	.074

Note: Dependent variable—Leadership: leading organization.

The results of the ANOVA showed a significant difference ($F_{3,112}$ = 50.585, p = .000) indicating that the relationship between the dependent variable of leadership: leading organization and the three independent variables of trust, knowledge management, and organizational performance are linear; therefore, the model significantly predicts the dependent variable of leadership: leading organization.

Table 6.2 shows the coefficients. The beta weights for the three independent variables of trust, knowledge management, and organizational performance are examined to determine the predictor variables that are most influential in predicting the dependent variable of leadership: leading organization. As can be seen, **trust and knowledge management** were the two predictor variables that were most influential in predicting **leadership: leading organization**.

The results of descriptive statistics are leadership: leading organization (Mean = 5.9461, STDV = .83001), trust (Mean = 6.1140, STDV = .70112), knowledge management (Mean = 6.1362, STDV = .71291), and organizational performance (Mean = 5.9754, STDV = .71844). The correlations are shown in Table 6.3.

RQ2: Which of the three predictor variables (trust, knowledge management, and organizational performance) are most influential in predicting leadership: leading people?

The test of multicollinearity confirmed the nonexistence of multicollinearity among the IVs. The tolerance level values for all IVs were above .1, and the VIF values for all IVs were greater than 10. The values were as follows: IV_1 = trust (TL = .474, VIF = 2.108), IV_2 = knowledge management (TL = .555, VIF =1.802), and IV_3 = organizational performance (TL = .519, VIF =1.926). Once the nonexistence of multicollinearity was established, we then interpreted (1) the model summary, (2) the ANOVA, and (3) the coefficients table.

The model summary revealed the results of multiple correlation (R = .737), squared multiple correlation (R^2 = .543), and the adjusted squared multiple correlation (R^2_{adj} = .531). These results indicated that the three independent variables

Table 6.3   Correlations

		LO	*T*	*KM*	*OP*
LO	Pearson correlation	1	.673[a]	.681[a]	.595[a]
	Sig. (1-tailed)		.000	.000	.000
	N	116	116	116	116
T	Pearson correlation	.673[a]	1	.629[a]	.659[a]
	Sig. (1-tailed)	.000		.000	.000
	N	116	116	116	116
KM	Pearson correlation	.681[a]	.629[a]	1	.582[a]
	Sig. (1-tailed)	.000	.000		.000
	N	116	116	116	116
OP	Pearson correlation	.595[a]	.659[a]	.582[a]	1
	Sig. (1-tailed)	.000	.000	.000	
	N	116	116	116	116

Note: LO = leadership: leading organization; T = trust; KM = knowledge management; OP = organizational performance.

[a] Correlation is significant at the 0.01 level (1-tailed).

of trust, knowledge management, and organizational performance can soundly predict the dependent variable of leadership: leading people.

The results of the ANOVA showed a significant difference ($F_{3,112}$ = 44.320, p = .000) indicating that the relationship between the dependent variable of leadership: leading people and the three independent variables of trust, knowledge management, and organizational performance are linear, therefore, the model significantly predicts the dependent variable of leadership: leading people.

Table 6.4 shows the coefficients. The beta weights for the three independent variables of trust, knowledge management, and organizational performance are examined to determine the predictor variables that are most influential in predicting the dependent variable of leadership: leading organization. As can be seen, **trust, knowledge management, and organizational performance** were the three predictor variables that were most influential in predicting **leadership: leading people**.

The results of descriptive statistics are leadership: leading people (Mean = 6.4586, STDV = .63247), trust (Mean = 6.1140, STDV = .70112), knowledge management (Mean = 6.1362, STDV = .71291), and organizational performance (Mean = 5.9754, STDV = .71844). The correlations are shown in Table 6.5.

Table 6.4   Coefficients

	Unstandardized Coefficients		Standardized Coefficients		
	B	Std. Error	Beta	t	Sig.
(Constant)	1.855	.402		4.620	.000
Trust	.281	.084	.312	3.363	**.001**
Knowledge management	.296	.076	.334	3.891	**.000**
Organizational performance	.178	.078	.203	2.285	**.024**

Note: Dependent variable—Leadership: leading people.

Table 6.5   Correlations

		LP	T	KM	OP
LP	Pearson correlation	1	.655[a]	.648[a]	.602[a]
	Sig. (1-tailed)		.000	.000	.000
	N	116	116	116	116
T	Pearson correlation	.655[a]	1	.629[a]	.659[a]
	Sig. (1-tailed)	.000		.000	.000
	N	116	116	116	116
KM	Pearson correlation	.648[a]	.629[a]	1	.582[a]
	Sig. (1-tailed)	.000	.000		.000
	N	116	116	116	116
OP	Pearson correlation	.602[a]	.659[a]	.582[a]	1
	Sig. (1-tailed)	.000	.000	.000	
	N	116	116	116	116

Note: LP = leadership: leading people; T = trust; KM = knowledge management; OP = organizational performance.

[a] Correlation is significant at the 0.01 level (1-tailed).

RQ3: Which of the three predictor variables (trust, knowledge management, and organizational performance) are most influential in predicting leadership: leading self?

The test of multicollinearity confirmed the nonexistence of multicollinearity among the IVs. The tolerance level values for all IVs were above .1, and the VIF values for all IVs were greater than 10. The values were as follows: IV_1 = trust (TL = .474, VIF = 2.108), IV_2 = knowledge management (TL = .555, VIF =1.802), and IV_3 = organizational performance (TL = .519, VIF =1.926). Once the nonexistence of multicollinearity was established, we then interpreted (1) the model summary, (2) the ANOVA, and (3) the coefficients table.

The model summary revealed the results of multiple correlation (R = .769), squared multiple correlation (R^2 = .591), and the adjusted squared multiple correlation (R^2_{adj} = .580). These results indicated that the three independent variables of trust, knowledge management, and organizational performance can soundly predict the dependent variable of leadership: leading self.

The results of the ANOVA showed a significant difference ($F_{3,112}$ = 53.936, p = .000) indicating that the relationship between the dependent variable of leadership: leading self and the three independent variables of trust, knowledge management, and organizational performance are linear, therefore, the model significantly predicts the dependent variable of leadership: leading self.

Table 6.6 shows the coefficients. The beta weights for the three independent variables of trust, knowledge management, and organizational performance are examined to determine the predictor variables that are most influential in predicting the dependent variable of leadership: leading organization. As can be seen, **trust and knowledge management** were the two predictor variables that were most influential in predicting **leadership: leading self**.

The results of descriptive statistics are leadership: leading people (Mean = 6.2522, STDV = .67847), trust (Mean = 6.1140, STDV = .70112), knowledge management (Mean = 6.1362, STDV = .71291), and organizational performance (Mean = 5.9754, STDV = .71844). The correlations are shown in Table 6.7.

Table 6.6 Coefficients

	Unstandardized Coefficients		Standardized Coefficients		
	B	Std. Error	Beta	t	Sig.
(Constant)	1.250	.407		3.068	.003
Trust	.530	.085	.548	6.246	.000
Knowledge management	.228	.077	.239	2.952	.004
Organizational performance	.060	.079	.064	.763	.447

Note: Dependent variable—Leadership: leading self.

Table 6.7   Correlations

		LS	T	KM	OP
LS	Pearson correlation	1	.741[a]	.621[a]	.565[a]
	Sig. (1-tailed)		.000	.000	.000
	N	116	116	116	116
T	Pearson correlation	.741[a]	1	.629[a]	.659[a]
	Sig. (1-tailed)	.000		.000	.000
	N	116	116	116	116
KM	Pearson correlation	.621[a]	.629[a]	1	.582[a]
	Sig. (1-tailed)	.000	.000		.000
	N	116	116	116	116
OP	Pearson correlation	.565[a]	.659[a]	.582[a]	1
	Sig. (1-tailed)	.000	.000	.000	
	N	116	116	116	116

Note: LS = leadership: leading self; T = trust; KM = knowledge management; OP = organizational performance.

[a] Correlation is significant at the 0.01 level (1-tailed).

Discussion

Koohang et al. (2017) proposed a research model that explored the leadership effectiveness influence on organizational trust among employees, knowledge management processes within organizations, and organizational performance. The findings of their study implied that effective leadership (i.e., leading organization, leading people, and leading self) in general, contributed to increased trust among employees, advanced the successful implementation of knowledge management processes, and in turn improved organizational performance.

The present study aimed to examine the influence of each leadership effectiveness (leading organization, leading people, and leading self) as a predicting variable on the three predictor variables of trust, knowledge management, and organizational performance within organizations. The leading organization encompasses the characteristics of a leader that assure organizational advancement. The leading people consists of the characteristics of a leader that enhance productivity among employees. The leading self holds the characteristics that augment the development of a leader.

The findings for leadership: Leading organization indicated that trust and knowledge management were the two predictor variables that were most influential in predicting leadership: Leading organization. In other words, the leading organization characteristics of a leader (i.e., change, innovation, influence, and diversity/inclusion) that collectively assure organizational advancement influence/predict (1) trust that ensures safe and satisfying relationships among people, and (2) knowledge management that secures effective and efficient handling of information and resources within organizations.

The findings for leadership: Leading people indicated that trust, knowledge management, and organizational performance were the three predictor variables that were most influential in predicting leadership: Leading people. In other words, the leading people characteristics of a leader (i.e., motivation, empowerment, listening, interpersonal communication, building relationship, and conflict) that collectively enhance productivity among employees influence/predict (1) trust that ensures safe and satisfying relationships among people, (2) knowledge management that secures effective and efficient handling of information and resources within organizations, and (3) organizational performance that is a tangible result/output (performance) of an organization as compared to its intended output.

The findings for leadership: Leading self indicated that trust and knowledge management were the two predictor variables that were most influential in predicting leadership: Leading self. In other words, the leading self characteristics (i.e., values/ principles, self-awareness, feedback, managing time, and learning) that augment development of a leader influence/predict (1) trust that ensures safe and satisfying relationships among people, and (2) knowledge management that secures effective and efficient handling of information and resources within organizations.

The findings of this study, in general, imply that effective leadership is the foundation of elevated trust, the successful implementation of knowledge management processes, and ultimately, enhanced organizational performance. In this chapter, we refer to effective leadership as characteristics that a leader must possess in leading organization, leading people, and leading self. These characteristics are not inherited, but learned and acquired by continuous leadership mentoring and development programs. Therefore, we recommend that leadership development programs include ongoing activities for employees on all the characteristics inherent to leading organization, leading people, and leading self.

In leading organization, the leadership development programs should include activities for leaders to learn

■ To be unconditionally open to change and innovation
■ To positively influence and set a vision so it can be translated into realistic business strategies and expected outcomes
■ To be respectful of diversity that brings about innovation and inclusion that guarantees the right conditions for everyone

In leading people, the leadership development programs should include activities for leaders to learn

- To continuously motivate employees to bring out the best in people
- To listen and empower people to do their jobs
- To demonstrate excellent interpersonal communication skills that become vital in bringing people together to work effectively
- To build and sustain relationships with subordinates; conflict resolution must be a part of development programs.

In leading self, the leadership development programs should include activities for leaders to learn

- To make decisions based on their values and principles
- To understand their strengths and weaknesses and how to be willing to improve
- To continuously seek feedback from others, to manage their time effectively, and always seek the opportunity to learn

Additionally, the development programs should include leadership development best practices such as 360-degree feedback, executive coaching, mentoring, networks, job assignments, and action learning.

Conclusion

Trust is vital because the presence of trust within organizations shape interpersonal relationships (Arrow, 1974), it develops and improves relationships (McLain & Hackman, 1999), and it is the foundation of any relationship that encourages and inspires decision-making within organizations (Porras, 2004). Knowledge management is vital because it is used to solve problems and make decisions within organizations (Davenport & Prusak, 1998). It influences performance and efficiencies within the organization (Drucker, 2000). The knowledge management processes ". . . expand, cultivate and apply available knowledge in ways that add value to the entity . . ." (Holsapple & Joshi, 2004, p. 596). Trust has been shown to have a positive link with knowledge management (Chenhall & Smith, 2003; Ferres, Connell, & Travaglione, 2005; Paliszkiewicz & Koohang, 2013; Paliszkiewicz, Koohang, Gołuchowski, & Nord, 2014) and organizational performance (Paliszkiewicz et al., 2014; Paliszkiewicz & Koohang, 2013). Ultimately, it is effective leadership that plays a vital role in the increased trust among people, the successful knowledge management processes, and the enhanced organizational performance. The leadership development programs and the leadership best practices must be a natural part of the organizational culture that values ongoing learning. Furthermore, these programs must be supported by all levels of leadership within organizations.

This study has limitations. A convenient sample was used with subjects participating in the study from only one geographic area in a country in Eastern Europe. To increase the generalizability of the results, we recommend that this study be expanded to include population samples from different countries and wider geographic areas.

References

Anderson, J. C., & Narus, J. A. (1990). A model of distributor firm and manufacturer firm working partnerships. *Journal of Marketing, 54*(1), 42–58.

Arrow, K. (1974). *The limits of organization.* New York, NY: Norton.

Awad, E. M., & Ghaziri, H. M. (2004). *Knowledge management.* New Jersey: Pearson Education International.

Barling, J., Weber, T., & Kelloway, E. K. (1996). Effects of transformational leadership training on attitudinal and financial outcomes: A field experiment. *Journal of Applied Psychology, 81*(6), 827–832.

Bass, B. M., (1960). *Leadership, psychology, and organizational behaviour.* New York, NY: Harper & Row.

Bassie, L. J. (1997). Harnessing the power of intellectual capital. *Training and Development, 51*(12), 25–30.

Brooking, A. (1999). *Corporate memory: Strategies for knowledge management.* London: International Thomson Business Press.

Charbonneau, D., Barling, J., & Kelloway, E. K. (2001). Transformational leadership and sports performance: The mediating role of intrinsic motivation. *Journal of Applied Social Psychology, 31*(7), 1521–1534.

Chenhall, R., & Smith, L. (2003). Performance measurement and reward systems, trust and strategic change. *Journal of Management Accounting Research, 15*(1), 117–143.

Chong, S. C., & Choi, Y. S. (2005). Critical factors of knowledge management implementation success. *Journal of Knowledge Management Practice, 6*(3), 21–37.

Chowdhury, S. (2005). The role of affect- and cognitions-based trust in complex knowledge sharing. *Journal of Managerial Issues, 17*(3), 310–326.

Costa, A. C. (2003). Work team trust and effectiveness. *Personnel Review, 32*, 605–622.

Costa, A. C., Roe, R. A., & Taillieu, T. (2001). Trust within teams, the relation with performance effectiveness. *European Journal of Work and Organisational Psychology, 10*, 225–244.

Darroch, J. (2003). Developing a measure of knowledge management behaviors and practices. *Journal of Knowledge Management, 7*(5), 41–54.

Das, T. K., & Teng, B. S. (2004). The risk-based view of trust: A conceptual framework. *Journal of Business and Psychology, 19*(1), 85–116.

Davenport, T. (1994). Saving IT's soul: Human centered information management. *Harvard Business Review, 72*(2), 119–131.

Davenport, T., & Prusak, L. (1998). *Working knowledge: How organizations manage what they know.* Boston, MA: Harvard Business School Press.

Debowski, S. (2006). *Knowledge management.* Milton: John Wiley & Sons Australia.

Doney, P. M., Cannon, J. P., & Mullen, M. R. (1998). Understanding the influence of national culture on the development of trust. *Academy of Management Review, 23*(3), 601–620.

De Toni, A., & Tonchia, S. (2001). Performance measurement systems-models, characteristics and measures. *International Journal of Operations & Production Management, 21*(1/2), 46–71.

Dore, R. (1983). Goodwill and the spirit of market capitalism. *British Journal of Sociology, 34*(4), 459–482.

Drucker, P. (2000). Knowledge-worker productivity: The biggest challenge. In J. W. Cortada & J. A. Woods (Eds.), *The knowledge management yearbook 2000–2001* (pp. 267–283). Woburn, MA: Butterworth-Heinemann.

Erikson, E. H. (1968). *Identity: Youth and crisis.* New York, NY: W. W. Norton Company.

Ferres, N., Connell, J., & Travaglione, A. (2005). The effect of future redeployment on organization trust. *Strategic Change, 14*(2), 77–91.

Foss, N. J., & Pedersen, T. (2002). Transferring knowledge in MNCs: The role of sources of subsidiary knowledge and organizational context. *Journal of International Management, 8*(1), 49–67.

Gandolfi, F., & Stone, S. (2016). Clarifying leadership: High-impact leaders in a time of leadership crisis. *Revista De Management Comparat International, 17*(3), 212–224.

Ganesan, S. (1994). Determinants of long-term orientation in buyer—Seller relationships. *Journal of Marketing, 58*(2), 1–19.

Garfinkel, H. (1967). *Studies in ethnomethodology.* Englewood Cliffs, NJ: Prentice Hall.

Gibb, J. R., (1978). *Trust, a new view of personal and organizational development.* Los Angeles, CA: Guild of Tutors Press, International College.

Gosselin, M. (2005). An empirical study of performance measurement in manufacturing organizations. *International Journal of Operations & Production Management, 54*(5/6), 419–437.

Grant, R. M. (1996). Toward a knowledge-based theory of the firm. *Strategic Management Journal, 17*(Special issue), 109–122.

Grojean, M. W., Resick, C. J., Dickson, M. W., & Smith, D. B. (2004). Leaders, values, and organizational climate: Examining leadership strategies for establishing an organizational climate regarding ethics. *Journal of Business Ethics, 55*(3), 223–241.

Grudzewski, W. M., Hejduk, I. K., Sankowska, A., & Wańtuchowicz, M. (2007). *Zarządzanie zaufaniem w organizacjach wirtualnych* [Trust management in virtual organization]. Warsaw: Difin.

Handy, C., (1995). Trust in virtual organization. *Harvard Business Review, 73*(3), 40–50.

Holsapple, C., & Joshi, K. (2004). A formal knowledge management ontology: Conduct, activities, resources, and influences. *Journal of the American Society for Information Science and Technology, 55*(7), 593–612.

Holsapple, C. W., Hsiao, S.-H., & Oh, J.-Y (2016). Parameters of knowledge management success. In J. Liebowitz (Ed.), *Successes and failures of knowledge management* (pp. 1–12). Cambridge, MA: Elsevier.

Horwitch, M., & Armacost, R. (2002). Helping knowledge management be all it can be. *Journal of Business Strategy, 23*(3), 26–32.

Howell, J. M., & Avolio, B. J. (1993). Transformational leadership, transactional leadership, locus of control, and support for innovation: Key predictors of consolidated-business-unit performance. *Journal of Applied Psychology, 78*(6), 891–902.

Jap, S. D. (1999). Pie-expansion efforts: Collaboration processes in buyer—Supplier relationships. *Journal of Marketing Research, 36*(4), 461–475.

Katz, D., & Kahn, R. L. (1966). *The social psychology of organizations.* New York, NY: Wiley.

Klimoski, R. J., & Karol, B. L. (1976). The impact of trust on creative problem solving groups. *Journal of Applied Psychology, 61,* 630–633.

Koohang, A., Paliszkiewicz, J., & Gołuchowski, J., (2017). The impact of leadership on trust, knowledge management, and organizational performance: A research model. *Industrial Management and Data Systems, 117*(3), 521–537.

Kouzes, J., & Posner, B. (2007). *The leadership challenge.* San Francisco, CA: Jossey-Bass.

Lee, K. C., Lee, S., & Kang, I. W. (2005). KMPI: Measuring knowledge management performance. *Information and Management,* (*42*)3, 469–482.

Lewis, J. D., & Weigert, A. (1985a). Trust as a social reality. *Social Forces, 63*(4), 967–985.

Lewis, J. D., & Weigert, A. J. (1985b). Social atomism, holism, and trust. *Sociological Quarterly, 26*(4), 455–471.

Lieberson, S., & O'Connor, J. F. (1972). Leadership and organizational performance: A study of large corporations. *American Sociological Review, 37*(2), 117–130.

Liebowitz, J. (Ed.). (2008). *Making cents out of knowledge management.* Scarecrow Press. Lanhman and Toronto.

Liebowitz, J. (Ed.). (2012). *Knowledge management handbook. Collaboration and social networking.* Boca Raton, FL/New York, NY: CRC Press/Taylor & Francis.

Liebowitz, J. (Ed.). (2016). *Successes and failures of knowledge management.* Cambridge MA: Morgan Kaufmann/Elsevier.

Mayer, D. M., Kuenzi, M., Greenbaum, R., Bardes, M., & Salvador, R. B. (2009). How low does ethical leadership flow? Test of a trickle-down model. *Organizational Behavior and Human Decision Processes, 108*(1), 1–13.

Mayer, R. C., Davis, J. H., & Schoorman, F. D. (1995). An integrative model of organization trust. *Academy of Management Review, 20*(3), 709–734.

McAllister, D. J. (1995). Affect and cognition based trust as foundations for interpersonal cooperation in organizations. *Academy of Management Journal, 38*(1), 24–59.

McInerney, C. (2012). Knowledge management and the dynamic nature of knowledge. *Journal of the American Society for Information Science and Technology, 53*(12), 1008–1016.

McLain, D. L., & Hackman, K. (1999). Trust, risk and decision-making in organizational change. *Public Administration Quarterly, 23*(2), 152–176.

Morgan, R. M., & Hunt, S. D. (1994). The commitment trust theory of relationship marketing. *Journal of Marketing, 58*(3), 20–38.

Neely, A., Gregory, M., & Platts, K. (2005). Performance measurement system design: A literature review and research agenda. *International Journal of Operations & Production Management, 25*(12), 1228–1263.

Nonaka, I., & Takeuchi, H., (1995). *The knowledge creating company.* New York, NY: Oxford University Press.

Paliszkiewicz, J. (2007). Knowledge management: An integrative view and empirical examination. *Cybernetics and Systems, 38*(8), 825–836.

Paliszkiewicz, J. (2013). *Zaufanie w zarządzaniu* [Trust in management]. Warsaw: Wydawnictwo Naukowe PWN.

Paliszkiewicz, J., Gołuchowski, J., & Koohang, A. (2015). Leadership, trust, and knowledge management in relation to organizational performance: Developing an instrument. *Online Journal of Applied Knowledge Management, 3*(2), 19–35.

Paliszkiewicz, J., & Koohang, A. (2013). Organizational trust as a foundation for knowledge sharing and its influence on organizational performance. *Online Journal of Applied Knowledge Management, 1*(2), 116–127.

Paliszkiewicz, J., Koohang, A., Goluchowski, J., & Nord, J. (2014). Management trust, organizational trust, and organizational performance: Advancing and measuring a theoretical model. *Management and Production Engineering Review, 5*(1), 32–41.

Paliszkiewicz, J., Koohang, A., & Horn Nord, J. (2014). Management trust, organizational trust, and organizational performance: Empirical validation of an instrument. *Online Journal of Applied Knowledge Management, 2*(1), 28–39.

Parmenter, D. (2009). *Key performance indicators: Developing, implementing, and using winning KPIs.* Hoboken, NJ: Wiley.

Porras, S. T. (2004). Trust as networking knowledge: Precedents from Australia. *Asia Pacific Journal of Management, 21*(3), 345–363.

Ring, P. S., & Van den Ven, A. H. (1992). Structuring cooperative relationships between organizations. *Strategic Management Journal, 13*(7), 483–498.

Rotter, J. B., (1967). A New scale for the measurement of interpersonal trust. *Journal of Personality, 35,* 651–665.

Sadler-Smith, E., Spicer, D. P., & Chaston, I. (2001). Learning orientations and growth in smaller organizations. *Long Range Planning, 34*(2), 139–158.

Schaubroeck, J., Hannah, S. T., Avolio, B. J., Kozlowski, S. W. J., Lord, R. L., Trevino, L. K.,... Pen, A. C. (2012). Embedding ethical leadership within and across organization levels. *Academy of Management Journal, 55*(5), 1053–1078.

Seppanen, R., Blomqvist, K., & Sundqvist, S. (2007). Measuring inter-organizational trust—A critical review of the empirical research in 1990–2003. *Industrial Marketing Management, 36,* 249–265.

Shapiro, S. P. (1987). The social control of impersonal trust. *American Journal of Sociology, 93*(3), 623–658.

Silva, A. (2016). What is leadership? *Journal of Business Studies Quarterly, 8*(1), 1–5.

Sinclair, D., & Zairi, M. (1995). Effective process management through performance measurement: Part II-benchmarking total quality-based performance measurement for best practice. *Business Process Management Journal, 1*(2), 58–72.

Sink, D., & Tuttle, T. (1989). *Planning and measurement in your organization of the future.* Norcross, GA: Industrial Engineering and Management Press.

Spender, J. C. (1996). Making knowledge the basis of a dynamic theory of the firm. *Strategic Management Journal, 17*(Special issue), 45–62.

Spender, J.-C., & Grant, R. M. (1996). Knowledge and the firm: Overview. *Strategic Management Journal, 17,* 5–9.

Stevens, J. (2001). *Applied multivariate statistics for the social sciences.* Hillsdale, NJ: Lawrence Erlbaum Associates.

Sveiby, K. E. (1987). *Managing knowhow.* London: Bloomsbury

Sztompka, P. (1999). *Trust: A sociological theory.* Cambridge: Cambridge University Press.

Teece, D. T. (1998). Capturing value from knowledge assets: The new economy, markets for know-how, and intangible assets. *California Management Review, 40*(3), 55–79.

Thomas, A. B. (1988). Does leadership make a difference to organizational performance? *Administrative Science Quarterly, 33*(3), 388–400.

Wang, S., and Noe, R. A. (2010). Knowledge sharing: A review and directions for future research. *Human Resource Management Review, 20,* 115–131.

White, G. P. (1996). A survey and taxonomy of strategy-related performance measures for manufacturing. *International Journal of Operations & Production Management, 16*(3), 42–61.

Winston, B. E., & Patterson, K. (2006). An integrative definition of leadership. *International Journal of Leadership Studies, 1*(2), 6–66.

Chapter 7

Trust and Knowledge Sharing: The Example of Higher Education in Poland

Barbara Kożuch and Regina Lenart-Gansiniec

Contents

Introduction

The potential of knowledge sharing in academia is greatly emphasized in the literature (Chua, 2003; Fullwood et al., 2013; Iqbal et al., 2011; Świgoń, 2015; Tippins,
2003). It is considered to be one of the factors for improving the quality, competi-
tiveness, and effectiveness of realizing research tasks and success in an increasingly
competitive environment (Alavi & Leidner, 1999; Syed-Ikhsan & Rowland, 2004;

Van den Hooff & De Ridder, 2004; Yang, 2007), and for building higher learning education institutions (Leja, 2013). The most recent literature indicates that knowledge sharing is dependent on organizational conditions. It is often said that a specific *sine qua* noncondition of knowledge sharing is trust. Moreover, trust increases the possibilities of building competitive advantage (Barney & Hansen, 1994; Ciancutti & Stending, 2001), raising the organization's effectiveness (Handy, 1995), reducing transactional costs, reacting to changes appearing in turbulent surroundings, and coordinating the organization (Shapiro, 1987). These phenomena may be investigated using the example of evaluating trust and knowledge sharing at the level of the organization as exemplified by higher education institutions in Poland. Knowledge about the level of sharing knowledge and trust in academia is important, as it may contribute to an efficient functioning of an educational institution and raising the quality of its offered services (Świgoń, 2015). A lack of this knowledge may lead to impeding the process of knowledge sharing and even losing valuable and difficult to achieve resources of the organization.

The aim of this chapter is to indicate the conditions of the level of knowledge sharing and the levels of trust. The choice of academia as the object of research is justified by the fact that knowledge sharing constitutes an important factor in these conditions, which enables, among others, the promotion of vocational skills and competencies of academic instructors (Huo, 2013; Semradova & Hubáčková, 2014).

This chapter presents the results of empirical research conducted during the period of January–February 2016. To obtain empirical data, a research tool was used in the form of a questionnaire directed to scholars employed at scientific and didactic posts. A total of 60 persons were examined.

This chapter is composed of three parts. The first presents the essence of trust and knowledge sharing. The second presents the existing research related to knowledge sharing and trust. The third and last part shows the results of the empirical research.

Theoretical Background

Trust

Trust is "an expression of confidence between parties in an exchange of some kind; confidence that they will not be harmed or put at risk by the actions of the other party or confidence that no party to the exchange will exploit the other's vulnerability" (Jones & George, 1998). It is considered the glue and lubricant of social capital (Anderson & Jack, 2002), the core of relationships (Mishra & Morrissey, 2000), since it leads to behaviors, which enable coping with risk and uncertainty connected with interaction. Trust constitutes a positive factor of collaboration; it concerns the expectations and predictability of behaviors of the trustee on the part of the trustor. It refers to both the convictions and the decisions and actions

(Dietz & Den Hartog, 2006). According to the opinion of Denise Rousseau et al., trust extends "from a calculated weighing of perceived gains and losses to an emotional response based on interpersonal attachment and identification" (Rousseau, Sitkin, Burt, & Camerer, 1998). In turn, Jo Paliszkiewicz interprets trust as the "bridge between past experiences and anticipated future" (Paliszkiewicz, 2011).

In the management literature, it points out the impact of trust on an organization's flexibility (Bibb & Kourdi, 2004), competitive advantage (Barney & Hanson, 1994; Ciancutti & Steding, 2001), development (Adler, 2001), collaboration (Mayer, Davis, & Schoorman, 1995), interaction and maintaining relations (Busacca & Castaldo, 2005), achieving business goals (Darrough, 2008), knowledge sharing (Gilbert & Tang, 1998), reduction of transactional costs and implementing changes in the organization (Hoey & Nault, 2002), employees' involvement (Busacca & Castaldo, 2005; Livet & Reynaud, 1998), and productiveness (Kramer & Cook, 2004), knowledge management (Castelfranchi, 2004), organizational learning (Bibb & Kourdi, 2004), free flow of information (Beccerra & Gupta, 1991), and organizational innovativeness (Sankowska, 2011).

Knowledge Sharing

Knowledge sharing is defined as a process through which knowledge possessed by an individual is transformed into a form which may be understood, acquired, and used by other individuals (Ipe, 2003). Knowledge sharing is a process that considers trial and error actions, feedback, constant adaptation of all the participants of this process, exploitation of existing knowledge to transfer and apply it for a better, faster, or less costly execution of a specific task compared to a situation in which knowledge sharing does not exist (Christensen, 2007).

From management's point of view, collective action consists of knowledge exchange within teams and organizational units, and organizations are at the core of knowledge sharing. The principal aim of knowledge sharing is to make use of available knowledge to improve the effectiveness of a given group of employees as well as to improve coordination of other processes in each organization (Alavi & Leidner, 1999; Salisbury, 2003).

Main Conditions of Trust and Knowledge Sharing

In the literature, it is acknowledged that trust is one of the most important elements of knowledge sharing (Davenport & Prusak, 1998). First, trust increases the probability that the members of the organization will willingly share knowledge and its use. This is important for building relations between members of the organization and the scope of sharing knowledge (McEvily, Perrone, & Zaheer, 2003). Second, trust eliminates the fears and suspicions that knowledge sharing is not mutually beneficial (Martin, 2004). Trust enables people to assess whether the benefits of sharing knowledge will be at least the same for each party (Luhmann, 1988), and at the same time this may reduce transaction costs (Nooteboom, 2003, p. 103). People trust others based on the condition that others will behave in a specific way (Mayer et al., 1995). Third, the conviction about the quality of knowledge and the competencies and credibility of both parties to knowledge sharing plays an important role in trust (Quigley, Tesluk, Locke, & Bartol, 2007). This increases the inclination to share personal knowledge (Chowdhury, 2005). Trust not only increases the volume of information that people share (Szulanski et al., 2004), but it also facilitates making decisions by simplifying the collection of information and its interpretation (McEvily et al., 2003). Moreover, trust is considered a factor that makes interactions between members of working parties and sharing knowledge possible (Tsai & Ghoshal, 1998).

As already mentioned, trust is considered to be one of the factors of knowledge sharing. In numerous research studies, it has been pointed out that it influences relations (Casimir et al., 2012), beliefs, and convictions of employees that the organization cares for their welfare (Renzl, 2008) and also stimulates employees to share knowledge of specific initiatives (Quigley et al., 2007). It also contributes to acceptance of new knowledge (Castelfranchi, 2004). Trust is considered as an individual factor of knowledge sharing, which means that trust and readiness to build relations with another person translates to readiness to share knowledge with others.

Method

The "web surveys" technique was applied. Higher education institutions with their seats in the territory of Poland were chosen as the object of research. The choice of an institution of higher education as the organization type was intentional and it resulted from the fact that on the one hand they play a specific role in knowledge sharing, while on the other hand they are trust organizations.

Research using the method of a diagnostic survey was carried out in the period between January and February 2016. To this aim, a survey questionnaire was used, which was placed on an Internet platform. The respondents within the realized research were academics experienced in working at a higher education institution, who had deep and expert knowledge related to problems of trust and knowledge sharing. They received electronic mail in their in-boxes, which was an invitation to participate in the research. The survey questionnaire was composed of seven questions. The respondents constituted a supplement. The respondents evaluated the situation present in an organizational unit, in which they are employed. On a 5-point scale, where "1" means "this sentence is totally false," and "5" means "this statement fully reflects the situation," they evaluated the atmosphere of trust present in the organization as well as knowledge sharing. The questionnaire was anonymous, which contributed to the return of a fully filled out sheet. Sixty questionnaires in total were received for the analysis. It was assumed that such sampling would be guaranteed by participation in the research of respondents who possessed a scope of knowledge about the studied phenomenon.

Among the institutions participating in the research, the clearly predominant ones had a dominant share of public ownership (48 academic institutions, which constituted 35.8% of the population), whereas private schools constituted 3.16% of the whole population of private higher education institutions in Poland (12 academic institutions). Sixty academics participated in the research. In the dominating part, 23 respondents held the academic title of professor. The same number was constituted by associate professors (23 persons). Whereas, the remaining group of respondents were doctoral students (7 persons) and assistant lecturers (7 persons). In the examined sample, men dominated and constituted 58.3% of the examined population—35 people, whereas there were 25 women. The biggest group of the examined persons were 34–40 years of age and from 41 to 50. The structure and number of the examined population did not allow for the drawing of generalized conclusions; however, it did constitute interesting material, which illustrated the analyzed processes. Taking this into consideration, the focus was on the opinions of all the examined population in the results' presentation. Only a part of the research results is presented in this chapter.

Analysis and Results

The problematic aspects of knowledge sharing in higher education institutions have been raised in numerous publications. However, the existing research has focused on identifying the barriers to sharing (Davenport & Cronin, 2000; Fullwood et al., 2013; Tippins, 2003; Wang & Noe, 2010) and the conditions for the effective sharing of knowledge (Kożuch & Lenart-Gansiniec, 2016). The aim of this research was to specify the level of knowledge sharing and trust. Realization of the aims was conducted in two stages: (1) identification of the level of knowledge sharing; and (2) identification of the level of trust.

It was assumed that an appropriate level of knowledge sharing may be dependent on several necessary conditions. Basing on literature review, the most important ones were specified and evaluated (Table 7.1).

The results obtained in the empirical research indicated that in the evaluated academic institutions the processes connected with knowledge are undervalued, especially those related to knowledge sharing. The respondents pointed out that the most important processes of the course of knowledge sharing in organizations are those connected with searching for information or the ability to locate persons who possess a specific type of knowledge. The low appraisal of the importance of the processes connected with a lack of the occurrence of rivalry between the employees, treating knowledge sharing as a priority and a strategic issue, the employees' knowledge of the systems and procedures of knowledge sharing, skills

Table 7.1 Conditions of the Level of Knowledge Sharing

Factors Influencing the Knowledge Sharing Process	Learning Level		
	Low	Moderate	High
	Answer (Number of Indications)	Answer (Number of Indications)	Answer (Number of Indications)
The employees know where to look for specific information.	0	2	58
At any time, information is available about the employees who have at their disposal specific knowledge resources, they may be contacted and make use of their assistance and consultation.	58	2	0
Workshops/meetings are held regularly within the organizational unit.	22	38	0
There is no rivalry between the employees caused by internal unfair competition.	60	0	0
The employees are constantly encouraged to solve problems in a team and to share knowledge with other employees.	10	40	10
Knowledge sharing is an issue that is an important part of the strategy realized by the organizational unit.	60	0	0
Most employees know the systems and procedures of knowledge sharing valid in the organizational unit.	60	0	0
Knowledge sharing is noticed and appreciated by the organizational unit's management.	0	58	2

Table 7.1 Conditions of the Level of Knowledge Sharing (*Continued*)

Factors Influencing the Knowledge Sharing Process	Learning Level		
	Low	Moderate	High
	Answer (Number of Indications)	Answer (Number of Indications)	Answer (Number of Indications)
Employees of the organizational unit possess the ability to conduct dialogue and discussion.	50	5	5
Employees of the organizational unit willingly talk to each other.	34	20	6
Great emphasis is put on mutual and common learning in the teams and groups.	60	0	0
Progress, results, and mistakes are systematically analyzed and discussed together.	60	0	0
Employees of the organizational unit are motivated to share knowledge by nonfinancial incentives.	22	18	20
During work in the organizational unit, the employee does not have time to share knowledge.	0	0	60
In the organizational unit, there are no assigned places for meetings and talks.	58	2	0
In the organizational unit, the employees know who to turn to in case of a lack of information on a given subject.	0	2	58

Source: Authors' own study conducted in 2016.

possessed by the employees in the scope of conducting discussions and dialogue, but also analyzing and discussing progress, results, and mistakes of the employees. One may attempt to state that, possibly, the causes are aspects that include a lack of assigned places for meetings and discussions and no time for sharing knowledge, as indicated by the respondents. A confirmation of such affairs are also the observations made by other researchers who analyzed the situation of knowledge sharing in the academic world in Poland (Białas & Wojnarowska, 2013; Świgoń, 2015). Similar barriers, based on literature analysis, have been indicated by Leja (2013).

It is worth emphasizing that knowledge is the academic institution's principal resource, whereas learning is a process that runs unceasingly. It should be underlined that the obtained answers enabled an evaluation of the level of knowledge sharing. The respondents were aware of the significance of learning, however, despite this they pointed out that individual actions are significant and not the collective ones. The lack of being interested in collaborating with others was indicated here. The motives and stimuli for collective learning were not observed.

In the case of knowledge sharing—it is not perceived as something important for the organization. The employees share knowledge in a spontaneous and unstructured way. There are specified procedures here. Low activity in the scope of knowledge sharing, skills only in the scope of identifying knowledge sources, or a lack of procedures may be a basis for ascertaining an unsatisfactory level of knowledge sharing. Even further, such actions may cause losing and obsoleting of the resources in the organization's possession, elongation of the knowledge management process, and organizational learning (Friesl, Sackmann, & Kremser, 2011).

The starting point in this part of the chapter is the assumption that knowledge sharing requires organizational trust. It is important to state here that trust is a complex and subjective notion. It manifests itself in various actions of the employees and managerial staff: starting from acting, through delegating authority or entitlements, independency of the employees, their participation in management or establishing cooperation, and ending with teamwork. Its level is dependent on many factors, among others the employees' experience, management style, readiness to cooperate, or a given person's predisposition (Stocki, Prokopowicz, & Żmuda, 2008). Its level is important in so far as it contributes to the readiness to cooperate (Shaw & Gaines, 1989), sharing knowledge (Gilbert & Tang, 1998), communication (Tschannen-Moran, 2001), and the employees' openness and readiness to accept changes (Czajkowska, 2008). What is important is that when there is a low level of trust or even a lack of trust, the employees' or managing staff's mistrust may contribute to a lack of involvement and fear of taking initiatives connected with knowledge (Zeffane & Connell, 2003).

The connections between the level of trust and knowledge are multidimensional. In the literature, it is pointed out that it has an impact on knowledge management (Kelloway & Barling, 2000), organizational learning (Loon Hoe, 2007), including knowledge flow (Beccerra & Gupta, 1999) and communicating

(Kohn, Maier, & Thalmann, 2008). Consequently, with the growth of the level of trust, the level of knowledge sharing increases (Sankowska, 2011). This dependency results from a will to cooperate, interact, or co-create, and participate.

Taking into consideration the connection between trust and the processes related to knowledge, for the needs of this chapter evaluation by the respondents of the level of trust in higher education institutions was important since the level of trust influences knowledge sharing (Table 7.2).

Table 7.2 Conditions of the Level of Trust

	Level of Trust		
	Low	Moderate	High
Factors Influencing the Level of Trust	Answer (Number of Indications)	Answer (Number of Indications)	Answer (Number of Indications)
In the organizational unit, it is common to put trust in others.	20	35	5
The management encourages the employees to participate in decision-making.	20	14	26
In the organizational unit, the employees express their opinions freely even if they differ from the opinions of the majority.	13	34	13
The employees of the organizational unit feel a positive bond with their direct superiors.	39	11	10
The employees in the organizational unit feel attached to other employees.	26	20	14

(Continued)

Table 7.2   Conditions of the Level of Trust (*Continued*)

	Level of Trust		
	Low	Moderate	High
Factors Influencing the Level of Trust	Answer (Number of Indications)	Answer (Number of Indications)	Answer (Number of Indications)
Mutual kindness of the employees has a positive effect on the results of the organizational unit's actions.	20	10	30
The employees of the organizational unit are open to changes and new ways of realizing actions.	20	14	26
In a general approach, the level of trust between the employees in the organizational unit is very high.	26	34	0

Source: Authors' own study conducted in 2016.

The results of research in the scope of the level of trust in higher education institutions indicated that the respondents emphasized encouraging the employees by the management to participate in decision-making and the impact of kindness on the results of an individual's actions. In contrast, the aspect connected with a positive bond of the employees with their direct superiors received the lowest evaluation result. On the one hand, the employees note the significance of mutual kindness and on the other hand, the low value of a positive bond with superiors. Such evaluation of the respondents points to their conviction about difficulties with trusting others. In addition, it should be emphasized that the trust level received a low (26 respondents) and moderate (34 respondents) evaluation result. None of the respondents stated that the level of trust in their academic world was at a very high level. The obtained research results correspond with the results of research by Świgoń (2015). The author emphasized here the issue of the specifics of working in a higher education institution's organizational units and the fact of work individualization and aversion to collective work. The reason for this is the evaluation performed by the management was based almost exclusively on scientific publications. To sum up, the level of trust was evaluated by the respondents as low or moderate. Considering the results of the research, the condition of organizational learning in the environments where the respondents functioned, did not confirm the regularity that there is a connection between trust and knowledge sharing as its part.

Discussion and Conclusion

The conducted research enabled the assessment of the level of knowledge sharing and trust related to the academic staff of higher education institutions having their seats in Poland. A few conclusions may be drawn on its basis.

1. The respondents assess the level of knowledge sharing in their workplace as low. They do not treat knowledge sharing as a priority; however, they did ascertain the importance of the processes connected with searching for information or the ability to locate people who possess a specific kind of knowledge. The obtained results correspond with the observations of other researchers of academia in Poland (Białas & Wojnarowska, 2013; Leja, 2013; Świgoń, 2015). Despite the awareness that knowledge sharing is important and it constitutes the basis for creating new ideas and developing new possibilities, it is not perceived by the respondents as important for them, their collaborators, and their working environment.
2. The level of trust was assessed by the respondents as low or moderate. They see the reason for this in weak ties between the employees and their direct superiors. Such assessment of the persons taking part in the research points to a conviction of the respondents about the existence of difficulty in placing one's confidence in others. The obtained results correspond with the research results by Świgoń (2015). The dominating individualization of work or rivalry contribute to avoiding knowledge sharing.
3. The deliberations presented in this chapter do not exhaust the problematic aspects raised. The conducted research may not constitute the basis for generalization related to a larger population. Nonetheless, the obtained results may justify the need for conducting further research, particularly on the problematic aspects of the impact of trust on knowledge sharing or the reasons for decreasing trust in academic working environments. Considering the above, knowledge about this may be valuable and it may constitute a premise for undertaking further research on knowledge sharing and trust.

References

Adler, P. (2001). Market, hierarchy, and trust: The knowledge economy and the future of capitalism. *Organization Science, 12*(2), 215–234.

Alavi, M., & Leidner, D. E. (1999). Knowledge management systems: Issues, challenges, and benefits. *Communications of Association of Information Systems, 1*(7), 1–37.

Anderson, A. R., & Jack, S. L. (2002). The articulation of social capital in entrepreneurial networks: A glue or a lubricant? *Entrepreneurship & Regional Development, 14*(3), 193–210.

Barney, J. B., & Hanson, M. H. (1994). Trustworthiness as a source of competitive advantage. *Long Range Planning, 4*(28), 175–190.

Beccerra, M., & Gupta, A. K. (1999). Trust within the organisation: Integrating the trust literature with agency theory and transaction cost economics. *Public Administration Quarterly, 23*(2), 177–203.

Białas, S., & Wojnarowska, M. (2013). Bariery dzielenia się wiedzą na uczelniach publicznych [Barriers to knowledge sharing on public universities]. *E-mentor, 1*(48), 56–59. Retrieved from http://www.e-mentor.edu.pl/artykul/index/numer/48/id/992.

Bibb, S., & Kourdi, J. (2004). *Trust matters for organizational and personal success* (pp. 121–131). New York, NY: Palgrave Macmillan.

Busacca, B., & Castaldo, S. (2005). Trust as a market-based resources: Economic value, antecedents and consequences. In K. Bijlsma-Frankema & R. K. Woolthuis (Eds.), *Trust under pressure* (pp. 149–169). Cheltenham: Edward Elgar.

Casimir, G., Lee, K., & Loon, M. (2012). Knowledge sharing: Influences of trust, commitment and cost. *Journal of Knowledge Management, 16*(5), 740–753.

Castelfranchi, C. (2004). Trust mediation in knowledge management and sharing. *Trust Management, 2995*, 304–318.

Chowdhury, S. (2005). The role of affect- and cognition-based trust in complex knowledge sharing. *Journal of Managerial Issues, 17*(3), 310–326.

Christensen, P. H. (2007). Knowledge sharing: Moving away from the obsession with best practices. *Journal of Knowledge Management, 1*(1), 36–47.

Chua, A. (2003). Knowledge sharing: A game people play. *Aslib Proceedings, 55*(3), 117–129.

Ciancutti, A., & Steding, T. L. (2001). *Built on trust: Gaining competitive advantage in any organization* (pp. 399–410). Chicago, IL: Contemporary Books.

Czajkowska, M. (2008). Kultura zaufania w organizacji. Istota—potrzeba—kształtowanie [Trust culture in an organization. The heart of the matter—the need—the shape]. *Annales Universitatis Mariae Curie-Skłodowska, Sectio H Oeconomia, 1*, 311–319.

Darrough, O. (2008). *Trust and commitment in organizations.* Saarbrücken: VDM Verlag Dr Muller.

Davenport, E., & Cronin, B. (2000). Knowledge management: Semantic drift or conceptual shift? *Journal of Education for Library and Information Science, 41*(4), 294–306.

Davenport, T. H., & Prusak, L. (1998). *Working knowledge: How organizations manage what they know.* Boston, MA: Harvard Business School Press.

Dietz, G., & Den Hartog, D. N. (2006). Measuring trust inside organisations. *Personnel Review, 35*(5), 557–588.

Friesl, M., Sackmann, S. A., & Kremser, S. (2011). Knowledge sharing in new organizational entities: The impact of hierarchy, organizational context, micro-politics and suspicion. *Cross Cultural Management: An International Journal, 18*(1), 71–86.

Fullwood, R., Rowley, J., & Delbridge, R. (2013). Knowledge sharing amongst academics in UK universities. *Journal of Knowledge Management, 17*(1), 123–136.

Gilbert, J., & Tang, T. (1998). An examination of organizational trust antecedents. *Public Personnel Management, 27*(3), 321–338.

Handy, C. (1995). Trust and the virtual organization. *Harvard Business Review, 73*, 40–50.

Hoey, J., & Nault, E. (2002). Trust: The missing ingredient in assessment. *International Journal of Engineering Education, 18*(2), 117–127.

Huo, M. (2013). Analysis of knowledge-sharing evolutionary game in university teacher team. *International Journal of Higher Education, 2*(1), 60–66.

Ipe, M. (2003). Knowledge sharing on organizations: A conceptual framework. *Human Resource Development Review, 2*(4), 337–359.

Iqbal, M. J., Rasli, A., Heng, L. H., Bin, M., Ali, B., Hassan, I., & Jolaee, A. (2011). Academic staff knowledge sharing intentions and university innovation capability. *African Journal of Business Management, 5*(27), 11051–11059.

Jones, G. R., & George, J. M. (1998). The experience and evolution of trust: Implications for cooperation and teamwork. *The Academy of Management Review, 23*(3), 531–546.

Kelloway, E. K., & Barling, J. (2000). What have we learned about developing transformational leaders. *Leadership and Organization Development Journal, 21*(1), 355–362.

Kohn, T., Maier, R., & Thalmann, S. (2008). Knowledge transfer with e-learning resources to developing countries: Barriers and adaptive solutions. In M. Breitner & G. Hoppe (Eds.), *E-Learning* (pp. 15–29). Heidelberg: Springer.

Kożuch, B., & Lenart-Gansiniec, R. (2016). Uwarunkowania skutecznego dzielenia się wiedzą na uczelni, [Effective knowledge sharing conditions at a university]. *Zarządzanie Publiczne. Zeszyty Naukowe Instytutu Spraw Publicznych Uniwersytetu Jagiellońskiego, 4*, 303–320.

Kramer, R. M., & Cook, K. S. (2004). *Trust and distrust in organizations: Dilemmas and approaches.* New York, NY: Russell Sage Foundation.

Leja, K. (2013). *Zarządzanie uczelnią. Koncepcje i współczesne wyzwania* [University Management. Concepts and contemporary challenges]. Warszawa: Wolters Kluwer.

Livet, P., & Reynaud, B. (1998). Organisational trust, learning and implicit commitment. In N. Lazaric & E. Loren (Eds.), *Trust and economic learning* (pp. 266–284). Londres: Edward Elgar.

Loon Hoe, S. (2007). Shared vision: A development tool for organizational learning. *Development and Learning in Organizations: An International Journal, 21*(4), 12–13.

Luhmann, N. (1988). Familiarity, confidence, trust: Problems and alternatives. In D. Gambetta (Ed.), *Trust: Making and breaking of cooperative relations* (pp. 94–107). Oxford: Blackwell.

Martin, W. J. (2004). Demonstrating knowledge value: A broader perspective on metrics. *Journal of Intellectual Capital, 5*(1), 77–91.

Mayer, R., Davis, J., & Schoorman, F. D. (1995). An integrative model of organizational trust. *Academy of Management Review, 20*(3), 709–734.

McEvily, B., Perrone, V., & Zaheer, A. (2003). Trust as an organizing principle. *Organization Science, 14*(1), 91–103.

Mishra, J., & Morrissey, M.A. (2000). Trust in employee/employer relationships: A survey of West Michigan managers. *Seidman Business Review, 6*(1), 14–15.

Nooteboom, B. (2003). Problems and solutions in knowledge transfer. In D. Fornahl & T. Brenner (Eds.), *Cooperation, networks and institutions in regional innovation systems* (pp. 105–125). Northampton: Edward Elgar.

Paliszkiewicz, J. (2011). Trust management: Literature review. *Management, 6*(4), 315–331.

Quigley, N. R., Tesluk, P. E., Locke, E. A., & Bartol, K. M. (2007). A multilevel investigation of the motivational mechanisms underlying knowledge sharing and performance. *Organization Science, 18*(1), 71–88.

Renzl, B. (2008). Trust in management and knowledge sharing: The mediating effects of fear and knowledge documentation. *Omega, 36*(2), 206–220.

Rousseau, R. M., Sitkin, S. M., Burt, R. S., & Camerer, C. (1998). Not so different after all: A cross-discipline view of trust. *Academy of Management Review, 23*(3), 393–404.

Salisbury, M. W. (2003). Putting theory into practice to build knowledge management systems. *Journal of Knowledge Management, 7*(2), 128–141.

Sankowska, A. (2011). *Wpływ zaufania na zarządzanie przedsiębiorstwem: Perspektywa wewnątrzorganizacyjna* [The impact of trust on the company's management: An intra-organisational perspective]. Warszawa: Difin.

Semradova, I., & Hubackova, S. (2014). Responsibilities and competences of a university teacher. *Procedia—Social and Behavioral Sciences, 159*, 437–441.

Shapiro, S. P. (1987). The social control of impersonal trust. *American Journal of Sociology, 9*(3), 623–658.

Shaw, M. L. G., & Gaines, B. R. (1989). A methodology for recognizing conflict, correspondence, consensus, and contrast in a knowledge-acquisition system. *Knowledge Acquisition, 1*(4), 341–363.

Stocki, R., Prokopowicz, P., & Żmuda, G. (2008). *Pełna partycypacja w zarządzaniu* [Full participation in management]. Kraków: Wolters Kluwer.

Świgoń, M. (2015). *Dzielenie się wiedzą i informacją: Specyfika nieformalnej komunikacji w polskim środowisku akademickim* [Knowledge and information sharing: Characteristics of polish informal scholarly communication]. Olsztyn: Wydawnictwo Uniwersytetu Warmińsko-Mazurskiego.

Syed-Ikhsan, S. O. S, & Rowland, F. (2004). Knowledge management in a public organization: a study on the relationship between organizational elements and the performance of knowledge transfer. *Journal of Knowledge Management, 8*(2), 95–111.

Szulanski, G., Cappetta, R., & Jensen, R. J. (2004). When and how trustworthiness matters: Knowledge transfer and teh moderating effect of causal ambiguity. *Organization Science, 15*(5), 600–613.

Tippins, M. J. (2003). Implementing knowledge management in academia: Teaching the teachers. *International Journal of Educational Management, 17*(7), 339–345.

Tsai, W., & Ghoshal, S. (1998). Social capital and value creation: The role of intrafirm networks. *Academy of Management Journal, 41*(4), 464–476.

Tschannen-Moran, M. (2001). Collaboration and the need for trust. *Journal of Educational Administration, 39*(4), 308–331.

Van den Hooff, B., & De Ridder, J. A. (2004). Knowledge sharing in context: The influence of organizational commitment, communication climate and CMC use on knowledge sharing. *Journal of Knowledge Management, 8*(6), 117–130.

Wang, S., & Noe, R. (2010). Knowledge sharing: A review and directions for future research. *Human Resource Management Review, 20*(2), 115–131.

Yang, J. (2007). The impact of knowledge sharing on organizational learning and effectiveness. *Journal of Knowledge Management, 11*(2), 83–90.

Zeffane, R., & Connell, J. (2003). Trust and HRM in the new millennium. *The International Journal of Human Resource Management, 14*(1), 3–11.

Chapter 8

Building Trust in Corporate Social Responsibility Reports

Jerzy Gołuchowski, Dorota Konieczna, and
Anna Losa-Jonczyk

Contents

Introduction

Companies strive to create good relationships with their stakeholders through a variety of means. Building the stakeholders' trust toward the companies and their products is one of the most important activities if they aim to properly function in the market and make a profit. One of the methods of building such trust includes socially responsible actions. However, the activities need to be in accordance with certain standards in order to actually build trust.

One of the challenges faced by a socially responsible company is the ability to build trust toward the company's environment (the brand or the product) in social reports (reporting on nonfinancial data). These reports play an important informative role for the stakeholders as well as orient the relations that the stakeholders shape with the organization. For the stakeholders, social reports fulfill a role that is similar to the role that financial reports fulfill for investors. Currently, nonfinancial data is increasingly more often reported in a single document along with the financial results. The reports also provide the organization managers with important knowledge allowing them to find the reference to the activities they perform.

Building and maintenance of trust in corporate social responsibility (CSR) reports is an area of reporting skills, good practices, procedures, and methods that may constitute a valuable subject for scientific research. The conducted literature research has proven that the undertaken analyses did not focus on the mechanisms associated with building trust in the reports; therefore, it was deemed valuable to conduct pilot studies on this subject.

In this chapter, an attempt has been to examine the methods of stakeholder trust development in selected reports created by companies associated with the energy industry. Here, trust is understood and analyzed as a relation based on three basic foundations: benevolence, integrity and predictability, and competence.

Building trust in the reports takes place by means of the natural language. In this study it is assumed, in accordance with the weak version of the Sapir-Whorf hypothesis, that language is not neutral and that a mere analysis of the report language allows, to a certain extent, an evaluation of the trust creation mechanisms (cf. Burgess, 2017). The language is reviewed from the discursive point of view and using discourse analysis methods. Discourse analysis is treated here as an eclectic method of text examination, which takes into account the context as well as the situation in which the given message was created. The socio-linguistic analyses are complemented with other studies necessary to create a proper description of the context and the economical background. Furthermore, the analysis is augmented with an assessment of the method of communication with the stakeholders, the context of the given company's functioning, and other necessary elements.

Study of Corporate Social Responsibility (CSR) Reports

The examined corporate social responsibility reports are prepared in accordance with standardized reporting method guidelines. The advantages of such standardization of social reports include the ability to quantify the quantitative data, the standardization of quality data, and the comparability of the results.

There are many international standards and norms associated with reporting on nonfinancial data as well as guidelines facilitating the execution of the reporting process available for organizations. The most important publications on this subject include the *Global Reporting Initiative* (GRI) *Guidelines* that contain rules and measures enabling the creation of a reporting system and the preparation of a report with quantified and comparable economic, environmental, and social results. Although the GRI guidelines are currently one of the most popular social reporting standards, they do not constitute the only source of knowledge on nonfinancial data reporting. The ISO norms such as: ISO 26000 (2017) (concerning the social responsibility of organizations CSR), ISO 14000 (2017) (concerning proenvironmental activities), ISO 14067 (2017) (concerning the carbon footprint), the SA8000 standard (2016), the OECD guidelines (2017), and a variety of industry guidelines (i.e., for the fuel, automotive, clothing, or finance industries) compose a wide range of documents supporting the preparation of nonfinancial reports.

The conducted literature research confirms that the social reports were studied, so far, that is, in the management sciences from the resource theory perspective referencing to the employee area as well as the organization's intellectual capital (Pedrini, 2007; Sułkowski, & Fijałkowska, 2013), in the context of communication with the stakeholders (Pasetti, Tenucii, Cinquini, & Frey, 2009; Knox, Maklan, & French, 2005), or in the context of tools for expanding the financial information reported by the companies to their stakeholders (Crisan & Zbuchea, 2015; Guthrie, Cuganesan, & Ward, 2007). Research on nonfinancial reporting was also conducted from the perspective of the stakeholder theory as well as company legitimacy (Harte & Owen, 1991; Oliveira, Rodrigues, & Craig, 2010; Vourvachis, Woodward, Woodward, & Patten, 2010).

The analysis of the literature on nonfinancial reporting did not prove that any study was undertaken in relation to the tools of trust building in the reporting process. Due to this fact, a theoretical and analytical model has been created to study the linguistic aspects of trust building in social reports and to analyze them in terms of the presentation of the CSR areas deemed in the ISO 26000 norm as a key for a socially responsible organization.

The ISO 26000 norm distinguishes seven main areas of socially responsible activities, which should be covered in corporate social responsibility reports.

The areas are as follows: organizational governance, labor practices, human rights, fair operating practices, consumer issues, community involvement and development, and environment. For the purposes of this chapter, the definitions of the CSR areas are taken from the ISO 26000 norm.

According to the ISO 26000 norm, **organizational governance** is a system by which the company makes and implements decisions in pursuit of its objectives. This includes both formal governance mechanisms based on the defined structures and processes, and informal ones resulting from the organization's culture and adopted values. Properly defined organizational governance enables the company to integrate social responsibility with the functioning throughout the organization and its relationships. It is a means of increasing the organization's ability to act in a socially responsible manner in relation to other core subjects/areas. An effective organizational governance system enables an effective inclusion of the social responsibility principles into the process of decision-making and decision implementation.

In the ISO 26000 norm, the **labor practices** area is associated with labor law practices where the company is regarded as an organization contributing to the improvement of the standards of living by means of ensuring safe employment and decent work. The company may undertake socially responsible activities through the realization of transparent employment practices, guaranteeing decent work conditions and social protection, ensuring employee development (e.g., through training), as well as through leading a social dialogue.

A socially responsible company is also obliged to respect **human rights**, according to the International Bill of Human Rights, also within the area of its influence. A socially responsible company should therefore exercise due diligence to identify, prevent, and address actual or potential human rights impacts resulting from the company's activities or the activities of other subjects the organization has relationships with.

Fair operating practices concern ethical conduct in the organization's dealings with other organizations (relationships with government agencies, business partners, suppliers, contractors, competition, and nongovernment organizations of which they are members). The issues of the social responsibility area described as "fair operating practices" include the activities associated with anti-corruption, responsible involvement in the public sphere, fair competition, promotion of social responsibility within the chain of values, and respect for property rights.

Activities associated with social responsibility undertaken within the area of **consumer issues** are fair marketing practices, protection of the consumer's health and safety, sustainable consumption, complaint and dispute resolution, consumer data protection and privacy, access to products and services, and education and catering for the needs of consumers that are vulnerable or in a disadvantaged social situation.

The ISO 26000 norm defines **community involvement and development** as a proactive attitude toward the local community with an aim to prevent and resolve social problems and foster partnerships with local organizations and stakeholders. A socially responsible company can, therefore, participate in and support civic institutions and support the civil participation processes in the determination of priorities related to social investments and activities aiming to develop the local community.

The subject of **environment** concerns the organization's activities limiting the negative impact of its presence on the natural environment. They may be associated with, among others, resource use, localization of the organization's activity, pollution emissions, waste production, and the influence of the organization on the natural habitat.

The manner of the reporting on the organization's activity in the seven abovementioned areas influences the process of building trust toward it, as it depicts how the organization understands its corporate social responsibility. Therefore, the analysis of CSR reports discern how organizations build trust toward themselves by undertaking selected CSR activities.

Theoretical Model of Trust Building in Social Reports

In order to develop the theoretical model of trust building in the reports as well as the tools used to study it, the authors used the trust concept presented/developed by Paliszkiewicz (in Chapter 5 in this book). The theoretical model takes into account the trust measures treated also as its foundations. Each foundation has been characterized using a list of selected features. The model is presented in Table 8.1.

According to the established theoretical model, one of the trust foundations is benevolence, which includes features such as the indication of interest, recognition of individual needs, and accessibility.

Indicating interest in CSR reports has been evaluated on three levels: (1) enumeration of the company stakeholders and provision of information on the manners of communication with them; (2) enabling the stakeholders to comment on the report; and (3) response to the raised remarks, engaging in the dialogue. Obviously, the dialogue itself can take different forms.

The recognition of individual needs should be based on consulting the particular elements of the report with an appropriate stakeholder group. The reports should therefore document such consultations as their quality influences building trust in stakeholders and report readers.

In the model it was assumed that accessibility is emphasized by two activities. The first one is associated with the selection of an appropriate genre in relation to the content as well as with the method of its presentation—whether the appropriate graphic, text, and multimedia form was selected.

The second activity is associated with the language complexity level. The level of the complexity of the language indicates that the creators of the report intend to be understood and through writing in a given, accessible manner they demonstrate respect toward the readers. The level of language complexity is evaluated according to the lists and solutions created for a given language. There are no universal solutions due to, among others, the type of the given language (e.g., measures prepared

Table 8.1 Theoretical Model of Trust Building in CSR Reports

Trust Foundation	Feature	Feature Justification
Benevolence	Interest indication	- Ensuring the ability to reply to a report proves the interest in the stakeholders and their opinion. - Interest in the stakeholders' remarks and response to their comments improve the manner of the company's functioning and are proof that the organization takes into account the opinions of the individuals interested/engaged in its activity.
	Individual needs recognition	- Information—included in the report—on the consultations conducted with the stakeholders in relation to the given areas is an indication of the recognition of the given stakeholder group needs.
	Accessibility	- The selection of an appropriate text genre, communication means, and a low text complexity level are a sign of respect toward the stakeholders and minimize the time needed to understand the message.
Integrity	Organization values and ethical behavior	- Value declarations and codes of ethics prove that the organization acts transparently and will not conduct any activity to the stakeholders' disadvantage.
	Action consistency and declaration realism	- The consistency of the chairpersons' declarations with the actions described in the reports enhances the company's credibility and proves its integrity.
Competence	Knowledge	- Clear determination on the scope of the company and what are the most important areas of its activity is decisive in terms of whether it will be perceived as a professional organization. - An expressed knowledge concerning the areas of interest of the given stakeholder groups proves the awareness of the significance of their actions for the benefit of the company.

Trust Foundation	Feature	Feature Justification
	Professionalism	- Professionalism and the ability to operate a business (expressed through, e.g., financial reports, information on the profit placed in an appropriate context of functioning in the market) is an indicator of the ability to properly function in the market.

Source: Authors' own elaboration based on Paliszkiewicz, J., *Zaufanie w zarządzaniu* [Trust in management], PWN, Warsaw, Poland, 2013.

for the isolation and positional English language are not appropriate for the inflected Polish language—cf. Majewicz, 1989). Therefore, when studying reports prepared in the Polish language, it is suggested to utilize the model created by the Plain Language Laboratory (PL: *Pracownia Prostego Języka*) at the University of Wroclaw (cf. Broda, Maziarz, Piekot, & Radziszewski, 2010; Piekot, Zarzeczny, & Moron, 2015). All models, regardless of the language for which they were created, are used to increase (or evaluate) the readability and comprehensibility of the text (cf. Piekot et al., 2015). A comprehensible text/report is one "that provides the average citizen with quick access to the information contained within it." A model that combines the evaluation of the text difficulty and its comprehensibility allows not only to indicate the text difficulty level but also to determine the preferred education of the reader of the given text. Therefore, when evaluating the text difficulty level, the Gunning text readability index was used (FOG = $0{,}4*[(LW/LZ) + 100*(LWT/LW)]$ where LW is the number of words within the text, LWT—number of words with more than 4 syllables (Broda et al., 2010), LZ—number of sentences, and the text comprehensibility tables which depend on the education and which were prepared by Broda et al. (2010) on the basis of the various Polish language corpora study. Data for the Gunning text readability index was obtained by means of feeding the report text into specially prepared software calculating the number of sentences, number of words within the sentences, and the number of syllables within a word. When speaking of the report text, we refer to the data depicted in the form of a written text. Any infographics, tables, charts, and so on, have been omitted. Furthermore, any statements presented in a form of a movie have not been analyzed.

Integrity (and predictability) is an important feature, on the foundation of which trust can be built. The study of integrity in social reports is based on the evaluation of the presentation of the organization's values and ethical behavior as well as on the consistency of actions and the realism of the declarations.

In their reports, companies usually declare the values and ethical behavior with which they align their business activity. If a company, aside from the declarations, has other documents related to its values (e.g., a code of ethics), it indicates a higher level of its maturity which, in turn, should induce increased trust. The confirmation that the declared values are in alignment with the described activities signifies an even higher maturity level.

Action consistency and declaration realism is another important integrity measure. If the declarations uttered in the CEO's letter are in accordance with the actual activities described in the report, this indicates a high level of the activities' compliance with the declarations and it therefore is a basis for trust building.

In the reports, **competence** can be indicated using a variety of methods. In the established theoretical model it was assumed that knowledge and activity professionalism are its most important discriminants. The level of knowledge was assessed on the basis of the organization's activity description, the method of its storage within the organization as well as the methods of its sharing. On the other hand, professionalism is understood as both good company operation (financial reports, profit, etc.) and properly prepared communication.

Methodology and Study Methods

Overall Study Description

In order to conduct the study on good practices of trust building in social reports, an analytical tool that evaluates the manner of trust building (and its particular dimensions) has been created on the basis of the developed theoretical model of the analysis of the linguistic aspects of building trust in such reports. Each of the three trust foundations has been allocated with features proclaiming the creation of the given trust dimension and the intensity of the given feature was assessed on the basis of measures characteristic for corporate social responsibility reports. A weight reflecting its influence on building trust in social reports was assigned to each feature (see Table 8.2). It was assumed that each foundation has the same weight.

Each of the reports has also been evaluated in terms of the dominating area of socially responsible activity (areas distinguished on the basis of the ISO 26000 norm). The particular reporting areas are determined as equally important, therefore the domination of a particular one indicates the activities deemed by the company as especially significant.

The results of the study on the methods of building trust have been compared with the analysis of the dominant reporting areas indicating the relationships between the activity area deemed as most important by the company and the strategy of trust building.

Analysis and Evaluation of Trust Foundations

Benevolence, in the mechanisms of trust building, has been evaluated by granting points for the presence of the particular indicators of the given feature within

Table 8.2 Tool Used for the Analysis of Trust Building Methods in the CSR Reports

Trust Foundation	Feature	Measure Confirming the Use of the Feature in the Report	Assessment Scale	Feature Weight
Benevolence	Interest indication	- Ensure the ability to reply to a report - Identification (if the ability has been ensured) of a response to the stakeholders' remarks	0–2	0,3
	Individual needs recognition	- A search for text references proving that the given report fragments were consulted with the appropriate stakeholder groups	0–1	0,3
	Accessibility	- Selection of the appropriate genre and means of communication (visual communication, infographics, etc.) - Evaluation of the language complexity level	0–2	0,4
Integrity	Organization values and ethical behavior	- Value declarations check - Verification whether the company is in possession of other documents regarding ethics: code of ethics, policies, and so on - Verification whether the declared values are compliant with the activities	0–3	0,5
	Action consistency and declaration realism	- Comparison of the declarations from chairpersons' letters with actual activities described in the report	0–2	0,5

(Continued)

Table 8.2 Tool Used for the Analysis of Trust Building Methods in the CSR Reports (*Continued*)

Trust Foundation	Feature	Measure Confirming the Use of the Feature in the Report	Assessment Scale	Feature Weight
Competence	Knowledge	- Knowledge on the company activities (information on this in the CSR report) - Knowledge regarding the stakeholders (if each of them will find something in the report)	0–2	0,4
	Professionalism	- Communication professionalism - Professionalism and the ability to operate a business (financial reports, profits, etc.)	0–3	0,6

Assessment values and their weights were established/selected using the expert evaluation method.

Source: Authors' own elaboration based on Paliszkiewicz, J., *Zaufanie w zarządzaniu* [Trust in management], PWN, Warsaw, Poland, 2013.

the limits of the established scale. In terms of interest indication, 1 point was granted for the enumeration of the stakeholders accompanied by the information on the communication means employed by the company in reference to them. One point was also granted for assurance that the stakeholders have the ability to comment on the report and for the response to their remarks (entering a dialogue). The recognition of individual needs means consulting the given report elements with an appropriate stakeholder group. One point was granted for the consultations of the particular elements with the stakeholder groups. The accessibility feature has been assessed by means of the appropriate indicators presence as well as their scale. The selection of the genre and form (multimedia, infographics, etc.) that is proper for the content—1 point and 1 point for the low level of language complexity according to the FOG index.

The analysis of integrity was divided into two main stages: the analysis of the organization's values and ethical behavior, and the analysis of the activity consistency and declaration realism. The analysis of the organization's values and ethical behavior was assessed on a scale of 0–3. One point was granted to an organization that declares, in the report, the values it follows in its business activity. Another point was granted if aside from this declaration, the company possesses other documents concerning the values, for example, code of ethics. The third point was granted if the declared values are in accordance with the described activities (if there is a relationship between the declared values and the activities). The analysis of the activity consistency is based on the comparison of the declarations found in the chairperson's letter with the actual activities covered in the report (full compliance was awarded with 2 points, partial compliance with 1 point).

The analysis of competence was conducted taking into account two features: knowledge and professionalism. Knowledge was evaluated in regard to the descriptions of the company's activities, what can be of interest for the particular stakeholder groups, and whether the information is available in the report. The company's activity description was assessed using a 0–2 scale (1 point was granted for the description presence and the second point for the content value, readability, and preparation related to the particular stakeholder groups).

Professionalism was assessed on the substantive matter of operating a business and its transparency. Making the financial reports, as well as other documents associated with operating a business, available was rewarded with 1 point. Making a profit or achieving a financial success (given the various macro- and microeconomic conditions in the particular industries, the indicators may differ) was worth 2 points.

Areas of Socially Responsible Activities

The ISO norm distinguishes seven areas of reporting (described in Chapter 9). The study of the areas' description method in the reports indicates the development direction of the company's image as a socially responsible organization.

The analysis of these areas determines how the company understands its responsibility and what tasks are identified as the most important to a socially responsible organization. The assessment of the most significant areas (from the company perspective) as well as of the information on the indicators reported in the given area (according to the GRI4 guideline) determines the organization's credibility. Indication of the report's dominating area(s) permits, in turn, to compare these areas with the company's activity scope (known from other documents and the company's functioning in the real world). This is significant due to the fact that there are organizations that treat CSR as a form of green washing of behavior that is not responsible in the field of core business activity.

Analysis of Trust in Selected CSR Reports

In order to verify the developed theoretical model as well as the analytical tool, pilot studies were conducted. The study material consisted of CSR reports published in 2016 (covering 2015 or joint reports for 2014 and 2015) created by five energy companies that participated in the Social Reports (2017) competition (PL: *Raporty Społeczne*) and whose reports are available on the http://raportyspoleczne. pl/ website, that is, Enea SA, Energa SA, EDF Polska, Tauron Polska Energia SA, and PGE Polska Grupa Energetyczna.

Each of the reports has been analyzed and evaluated in terms of the report creation standards as well as the methods of building trust in business, based on the three trust foundations assumed in the theoretical model (Paliszkiewicz, 2013). Due to space limitations, this chapter describes the analysis of a single report—created by Enea SA; however, it also contains the remaining companies' results.

Benevolence of Trust in a Selected Company's Report

In the Enea SA 2015 report, 1 point has been granted in terms of interest indication. The report contains the stakeholder information—"The map of Enea Group stakeholders" (PL: *Mapa interesariuszy Grupy Enea*) complemented by the information on the communication channels between the given group and the management as well as between the stakeholders themselves (although it is worth mentioning that the communication channels also include elements where one-sided management-stakeholders communication is dominant, e.g., annual report). The creators of the CSR report did not foresee the possibility of an open discussion on the content of the report. A contact regarding the report is possible only through the Public Relations and Communication Department of the Corporate Social Responsibility Office (PL: *Departament Public Relations i Komunikacji przy Biurze Społecznej Odpowiedzialności Biznesu*)—the company email address is available. Moreover, the customers may contact the company via Facebook (Energia+), where the company employees are engaged in discussions, although none of the posts were associated with the CSR report.

The recognition of individual needs is based, in the analyzed report, mainly on consulting the given areas of the report with the respective stakeholder groups. The report mentions that the key corporate values were established in agreement with the employees: "In collaboration with our staff, we have developed four key corporate values on which we base our business" (PL: *Wspólnie z naszymi pracownikami opracowaliśmy cztery kluczowe wartości korporacyjne, na których opieramy naszą działalność*). Furthermore, the originator and initiator of the Noble Box project, pr. Jacek Stryczka, was also invited into the discussion on ethical matters. What is more, the company conducts periodic customer satisfaction surveys (both individual and business-oriented). The company declares that the customer's opinions translate into actual changes and company actions. The report does not mention any declarations associated with the consultations of other report elements with the stakeholders. Therefore, the individual that needs identifications was awarded with 0,5 points.

The Enea Group's accessibility has been assessed for 1 point. In the first stage—the selection of an appropriate genre for the content and the method of its presentation—the attention was brought to the introduction of multimedia elements, infographics, tables, and charts. The report begins with a schematic animation accompanied by a short text information. The combination of these elements shows, in a simple way, what the Enea Group activity focuses on and what are its most important values. The report introduction page also features a short documentary on the creation of a modern coal power plant. The important sections are marked with both text and icons. Also, the report recipient has access to interactive data displayed in tables and charts (the reader can choose the presentation form) where they can change the time. However, the report creators allow only five factors to be used in this form. Furthermore, the report features several films, infographics, graphs, charts, and tables. These elements facilitate the reception and analysis of the covered content. The more important report elements (including means of contact with particular departments) have been highlighted with a light blue text background. A guiding idea has been assigned to each section and it is visible on the main banner. The company was awarded with 0,75 points for this element.

The text complexity level has been determined using the Gunning text readability index, which is equal to 17,5 for the Enea report. For this element, the company was granted 0,25 points.

Analysis of Integrity in a Selected Company's Report

In the Enea SA CSR report, a separate *Ethical Standards* subsection under the Enea Group section was devoted to the company's values. Following the consultations, the most important values determined by various stakeholder groups were: integrity, competence, responsibility, and safety. The values also appear in the presentation in the beginning of the report. Furthermore, the company has created the *Enea Capital Group Code of Ethics* and each new employee undergoes training on company values. There are also e-learning courses available on the intranet. The organization has established a code of ethics commission that can be contacted via email or through a special contact form.

The report highlights the activities aiming to improve both the safety of the employees during their work (through a series of various measures resulting from the law regulations and additional activities), the safety of the energy delivery to the customers and the broadly understood energy security. The remaining values: integrity, competence, and responsibility are not emphasized in such a direct manner; however, their importance can be observed in, for example, the declarations associated with the employees' engagement in education (competence sharing) or the high complexity level of the newly build batteries (and their service), and so on. Therefore, after an analysis of the organization values and ethical behavior, the company was granted 3 points.

Action consistency and declaration realism was assessed for 1,5 points. In the letter opening the report, the Enea Group CEO declares that the company is responsible for "both (. . .) the Employees, local communities, the quality of life of millions of Customers and the energy security of the country" (PL: *zarówno za (. . .) Pracowników i społeczności lokalne, jakość życia milionów Klientów, jak i za bezpieczeństwo energetyczne kraju*). The actions associated with the country's energy security, according to the chairperson, include the construction of a power unit in the Kozienice power plant. Furthermore, staff development is a priority, however, the letter does not feature any examples of such actions. On the other hand, it does feature the employees' activities for the benefit of the local environment. The comparison of the letter declarations with the entire report indicates the fulfilment of their majority.

Analysis of Competence in a Selected Company's Report

The knowledge indicator associated with the company's activity description in the Enea report was assessed for 2 points. The description in the report is complete and accompanied by multimedia material as well as infographics explaining the operation of the most important systems in the Enea Group.

Professionalism was assessed for 2 points.

Trust in Analyzed Reports of Companies in the Energy Industry

A similar analysis of corporate social responsibility reports was conducted in relation to the remaining four companies from the energy industry. Table 8.3 summarizes the numerical results of the analyses of all the organizations.

Tauron and PGE Polska Grupa Energetyczna utilize the corporate social responsibility report best in terms of building trust. Both companies prefer simple communication—the language complexity level is much lower than in the remaining

Table 8.3 Results of the Analysis of Trust Building in the Corporate Social Responsibility Reports

Company	Trust Foundation	Feature	Evaluation	Evaluation per Weight	Total (per Weight)
Enea SA	Benevolence	Interest indication	1	0,85	5,1
		Individual needs recognition	0,5		
		Accessibility	1		
	Integrity	Organization values and ethical behavior	3	2,25	
		Action consistency and declaration realism	1,5		
	Competence	Knowledge	2	2	
		Professionalism	2		
Energa SA	Benevolence	Interest indication	0,5	0,425	4,125
		Individual needs recognition	0,25		
		Accessibility	0,5		
	Integrity	Organization values and ethical behavior	2,5	2	
		Action consistency and declaration realism	1,5		
	Competence	Knowledge	2	1,7	
		Professionalism	1,5		

(Continued)

Table 8.3 Results of the Analysis of Trust Building in the Corporate Social Responsibility Reports (Continued)

Company	Trust Foundation	Feature	Evaluation	Evaluation per Weight	Total (per Weight)
Tauron Polska	Benevolence	Interest indication	1,5	1,4	6
		Individual needs recognition	0,5		
		Accessibility	2		
	Integrity	Organization values and ethical behavior	2	2	
		Action consistency and declaration realism	2		
	Competence	Knowledge	2	2,6	
		Professionalism	3		
PGE Polska Grupa Energetyczna	Benevolence	Interest indication	1,75	1,4	6
		Individual needs recognition	0,25		
		Accessibility	2		
	Integrity	Organization values and ethical behavior	2,5	2	
		Action consistency and declaration realism	1,5		
	Competence	Knowledge	2	2,6	
		Professionalism	3		

Table 8.3 Results of the Analysis of Trust Building in the Corporate Social Responsibility Reports *(Continued)*

Company	Trust Foundation	Feature	Evaluation	Evaluation per Weight	Total (per Weight)
EDF Polska	Benevolence	Interest indication	1	0,475	3,475
		Individual needs recognition	0,25		
		Accessibility	0,25		
	Integrity	Organization values and ethical behavior	2	1,5	
		Action consistency and declaration realism	1		
	Competence	Knowledge	1,5	1,5	
		Professionalism	1,5		

Source: Authors' own elaboration based on Paliszkiewicz, J., *Zaufanie w zarządzaniu* [Trust in management], PWN, Warsaw, Poland, 2013.

reports. Both companies used clear forms in order to communicate complicated financial data (as well as other details), that is, infographics, films, multimedia, and various charts. What is more, both companies achieved the highest indices of knowledge and professionalism. In turn, these features—constituting the benevolence and competence trust foundations—were decisive in terms of the predominance of their reports. All of the analyzed reports had a high integrity index.

Analysis of CSR Areas in Reports of Selected Polish Energy Companies

The CSR reports are evaluated in relation to seven areas distinguished in the ISO 26000 norm. These include: organizational governance, labor practices, human rights, fair operating practices, consumer issues, community involvement and development, and environment. For the purpose of this study, an evaluation of the most broadly described area was conducted. This has been completed by means of verification of the number of GRI4 indicators reported in the given area. None of the companies was obliged to report on the given area in accordance with all indicators.

Analysis of a Selected Company's Report

"Environment" was the dominant area in the report created by the Enea Group (Table 8.4). The company reported a total of 13 indicators associated with ecology and the influence of the organization's activities on the natural environment. The highest number of indicators reported in the "environment" area may result from the characteristics of the energy industry, which is largely based on the utilization of natural resources and where the energy production and distribution processes are associated with a high impact on the environment. Another

Table 8.4 Indicators in the CSR Areas Reported by Enea

Area	Organizational Governance	Labor Practices	Human Rights	Fair Operating Practices	Environment	Consumer Issues	Community Involvement and Development
Number of reported indicators	9	4	0	4	13	4	6

Source: Authors' own elaboration based on Paliszkiewicz, J., *Zaufanie w zarządzaniu* [Trust in management], PWN, Warsaw, Poland, 2013.

widely reported area covered by the Enea Group CSR report was "organizational governance" (9 indicators). In the years 2014–2015, the group has completed the "Corporate Governance—Enea Capital Group Management Plan" project (PL: *Ład korporacyjny—Plan Zarządzania Grupą Kapitałową Enea*), which aimed to simplify the decision-making processes, shorten the information flow, and concentrate on the Group's core activity.

Four indicators that are present in the report occur in the following areas: labor practices, fair operating practices, and consumer issues. The amount of information on key group stakeholders: employees, contractors, and customers seems to be low, given the fact that the Group's strategic objectives presented in the report refer to building long-lasting relationships with the customers and the optimal use of the organization's potential. A higher number of indicators (6) was present in activities related to the cooperation with the community. The report did not feature any indicators related to the "human rights" area.

Analysis of CSR Areas in Reports of Analyzed Energy Industry Companies

The analysis of the nonfinancial reports created by the Enea Group, as well as the remaining energy companies, proves that the activities associated with the company's influence on the environment constitute the most widely reported area in most of the discussed cases (Table 8.5). In the reports of three out of five analyzed companies, the "environment" area was described with 13 indicators (the highest

Table 8.5   Indicators in CSR Areas in Particular Energy Company Reports

| Companies | Number of Indicators Reported in the Given CSR Areas | | | | | | |
	Organizational Governance	Labor Practices	Human Rights	Fair Operating Practices	Environment	Consumer Issues	Community Involvement and Development
PGE	7	11	2	3	10	7	4
Tauron	7	11	2	1	17	18	2
EDF	9	4	0	4	13	4	6
Energa	5	9	0	1	13	5	3
Enea	9	4	0	4	13	4	6

Source: Authors' own elaboration based on Paliszkiewicz, J., *Zaufanie w zarządzaniu* [Trust in management], PWN, Warsaw, Poland, 2013.

number of indicators in comparison to the other areas). The remaining two companies (PGE and Tauron) reported 11 and 18 indicators in this area, which also signifies a large interest in this field of social responsibility. The number of indicators was lower by one in comparison to the number of indicators in the areas most broadly reported by the analyzed companies (for PGE it was "labor practices" with 11 indicators; for Tauron it was "consumer issues" with 18 indicators).

The studied companies also deemed labor practices, organizational governance, and consumer issues as significant areas of social responsibility (areas with a high number of indicators). The lowest number of indicators was associated with human rights and fair operating practices.

Conclusion

In this chapter, the theoretical model was presented as well as the tool for the analysis of the methods of trust building in the reports. To verify the proposed approach, social reports created by selected companies from the energy industry were analyzed. The undertaken study indicates the way in which the enterprise may build the stakeholder's trust toward it using the communication method known as the social responsibility report. The form of the report has been reviewed as an element that not only communicates the company's activity but may also be used to build trust.

The analysis of the language features of the corporate social responsibility reports and the nonlingual conditions of the company's functioning in the economic space, including the analysis of the dominant reporting areas, permitted us to distinguish the mechanisms used to build trust toward the company. The conducted pilot study shows that the study tool proposed by the authors of this chapter enables the evaluation of the corporate social responsibility reports in terms of trust building. Furthermore, evidence shows that elements such as the complexity level of the language used in the communication with stakeholders, the indication of ethical principles employed by the company, consistency or compliance of the declarations with real actions are considered significant elements of building such trust through the CSR reports, similar to clear information on the organization's financial results.

The conducted study and the presented report analysis results should be considered as pilot studies. The subsequent step includes the verification of the developed analytical tool on the basis of the discussions with the project authors as well as consultants in the field of social responsibility. The conclusions resulting from the study should serve as outlines for further studies on the tools associated with building trust toward an organization.

References

Broda, B., Maziarz, M., Piekot, T., & Radziszewski, A. (2010). Trudność tekstów o Funduszach Europejskich w świetle miar statystycznych [The difficulty of texts on the European Funds in light of the statistical measures]. In J. Miodek & W. Wysoczański (Eds.), *Rozprawy komisji językowej* [Dissertations of the language committee]. Wrocław, Poland: WTN.

Burgess, F. (2017, January 16). *The language of corporate governance: A sociological analysis.* Retrieved from http://www.lccge.bbk.ac.uk/publications-and-resources/postgraduate-research/docs/130930-Dissertation-Language-of-CorpGov.pdf

Crisan, C., & Zbuchea, A. (2015). CSR and social media: Could online repositories become regulatory tools for CSR related activitie's reporting? In A. Adi, G. Grigore, & D. Crowter (Eds.), *Developments in corporate governance and responsibility* (pp. 197–219). Bingley: Emerald Group Publishing.

Enea CSR Report. (2015). Retrieved from http://raportcsr.enea.pl/2015/pl/list-prezesa-0

Global Reporting Initiative Guidelines. (2017, January 16). Retrieved from https://www.globalreporting.org

Guthrie, J., Cuganesan, S., & Ward, L. (2007). Extended performance reporting: Evaluating corporate social responsibility and intellectual capital management. *Issues in Social and Environmental Accounting, 1*(1), 1–25.

Harte, G., & Owen, D. (1991). Environmental disclosure in the annual reports of British companies: A research note. *Accounting, Auditing and Sustainability Journal, 4*(3), 51–61.

ISO 14000. (2017, January 16). Family–Environmental Management. Retrieved from www.iso.org/iso/iso14000

ISO 14067. (2017, January 16). ISO/TS 14067:2013. Retrieved from www.iso.org/iso/catalogue_detail? csnumber=59521

ISO 26000. (2017, January 16). Discovering ISO 26000. Retrieved from www.pkn.pl/sites/default/files/discovering_iso_26000.pdf

Knox, S., Maklan, S., & French, P. (2005). Corporate social responsibility: Exploring stakeholder relationships and programme reporting across leading FTSE companies. *Journal of Business Ethics, 61*, 7–28.

Majewicz, A. F. (1989). *Języki świata i ich klasyfikowanie* [Languages of the world and their classification]. Warsaw, Poland: PWN.

Oliveira, L., Rodrigues, L. L., & Craig, I. (2010). Intellectual capital reporting in sustainability reports. *Journal of Intellectual Capital, 11*(4), 575–594.

OECD Guidelines. (2017, January 16). Overview OECD Guidelines for Multinational Enterprises. Retrieved from https://www.oecd.org/corporate/mne/38111315.pdf

Paliszkiewicz, J. (2013). *Zaufanie w zarządzaniu* [Trust in management]. Warsaw, Poland: PWN.

Pasetti, E., Tenucii, A., Cinquini, L., & Frey, M. (2016, December 1). *Intellectual capital communications: Evidence from social and sustainability reporting.* MPRA Munich Personal RePEc Archive. Retrieved from https://mpra.ub.uni-muenchen.de/16589/1/MPRA_paper_16589.pdf

Pedrini, M. (2007). Human capital convergences in intellectual capital and sustainability reports. *Journal of Intellectual Capital, 8*(2), 346–354.

Piekot, T., Zarzeczny, G., & Moron, E. (2015). Upraszczanie tekstu użytkowego jako (współ) działanie. Perspektywa prostej polszczyzny [Simplification of a non-literary text as a (co)operation. The plain Polish perspective]. In S. Niebrzegowska-Bartmińska, M. Nowosad-Bakalarczyk, & T. Piekot (Eds.), *Działania na tekście. Przekład—redagowanie—ilustrowanie* [Text operations. Translation—editing—illustration]. Lublin, Poland: Wyd. UMCS.

Social Reports (2017, January 16). Social reports. Retrieved from http://raportyspoleczne.pl

SA8000 Standard (2016, December 1). Social Accountability International. Retrieved from www.sa-intl.org/sa8000

Sułkowski, Ł., & Fijałkowska, J. (2013). Corporate social responsibility and intellectual capital interaction and voluntary disclosure. *Studia Ekonomiczne, Zeszyty Naukowe Uniwersytetu Ekonomicznego w Katowicach, 150*, 150–157.

Vourvachis, P., Woodward, T., Woodward, D. G., & Patten, D. M. (2010). CSR disclosure in response to major airline accidents: A legitimacy-based exploration. *Sustainability Accounting, Management and Policy Journal, 7*, 26–43.

Chapter 9

Trust and Marketing

Katarína Fichnová, Łukasz P. Wojciechowski,
and Peter Mikuláš

Contents

Introduction and Aims of the Chapter

Nowadays, marketing communication experts face the difficult challenge of how to attract the attention of customers. Conventional marketing communication is in a crisis (Saucet, 2015). Advertising expenses in past years have been rising in European Union (EU) countries (see IAB Europe Report, 2016).

With regard to the decreasing tendency of classical marketing forms, the creative approach (with originality as the main component) is becoming necessary. It doesn't only attract attention, but also helps to create a positive attitude toward a presented product or service, acceptance and memorability of a message, which helps to establish trust. The aim of this chapter is to explain these facts. At the same time, we want to illustrate the significant differences by using original research results and analyze the intersections between the two different groups of generations. Our results are also applicable in the marketing practice. The research demonstrates who and what Generations X and Y trust, how the various advertising types placed in different media are evaluated, and whether the creative execution can have an impact on credibility.

Review of the Literature

Trust in Marketing Communication

Marketing communication is aimed at the stimulation and regulation of the needs that are fulfilled by the means of purchase of the product or service (Pavlů, 2013). Trust should become the key entity of all communication measures and components of the communication mix.

Nevertheless, within the well-known models of marketing communication, the element of trust is included merely in the so-called model of "balloon effect" proposed by Lutz, MacKenzie, and Belch (1983), where trust represents one of the main components of "central elaboration of the advertising message" (in the *Elaboration Likelihood Model*, by Petty & Cacioppo, 1986)—where these factors are present: credibility of the source, ads, and concrete advertising—that is, a realization. In this regard, the credibility of advertising means whether the target group of people perceives the advertised product or brand as credible and believable.

The perceived congruence between the advertising message and the product (or brand) perception from the consumer perspective, determined mainly by his previous experience, is thus crucial for the credibility of advertising or another form of marketing communication. Therefore, the perceived credibility is the part of the process of creating and forming the attitude toward advertising based on rational analysis, which represents the central way of attitude formation. At the same time, it is supposed (compared with Světlík, 2012) that the receiver of the message also evaluates the credibility of the company producing the goods, communications medium, and previous and current advertising activities of the source. In the so-called peripheral elaboration of the advertising message, is trust included only implicitly?

Regarding the complexity of trust in marketing and communications, the model synthesizing all its essential and important components would be desirable for a better understanding. Various authors have conceptualized such a model. Garbarino and Johnson (1999) have stated that the integrative model of trust in marketing communication will vary by customer type (and thus, in continuum from transactional to highly relational bonds). Doney, Cannon, and Mullen (1998) confirmed that the integrative model must respect national cultures. Social norms and values influence the trust building in the context of business. The trust building between the seller and customer is essential and useful for both sides because trust can lower transaction costs, facilitate inter-organizational relationships, and enhance relationships. Advertising, however, generally does not have a very good reputation among consumers. In 1974, in a research study by Haller (1974), 75% of respondents believed that invalid or misleading statements came from half of all advertising. However, even current conventional marketing communication is in crisis (see Saucet, 2015). There is an increase in resistance to advertising, negative attitudes, and efforts to avoid it. Smith (2006, p. 21) from Yankelovich Partners Inc. (realized in 2004) stated that 60% of the U.S. population had more negative opinions on advertising and marketing than a few years before.

The absence of trust in advertising is reflected in the degree of skepticism on the truth and the overall content of advertising messages. Intercultural differences have been observed in this area—Dichl, Mueller, and Terlutter (2007) compared the sample from Germany and the United States, whereby it was proven that, in general, U.S. customers are less skeptical toward ads than German customers. The level of trust also varies depending on the advertising medium. Research by Soh, Reid, and King (2007) proved that Internet advertising is the least reliable compared with TV, radio, or print ads. The most trusted were magazines. The level of trust in advertising placed in different media is also influenced by the customers' income. The respondents with higher incomes trust magazines more, and trust the Internet less. While respondents with less income, trust ads on TV more and trust magazines less, and Internet ads are the second-most important for them (compared with TV) (Soh et al., 2007). The trust is also impacted by education—those who are more educated, trust significantly less advertising in media. To enhance the credibility of the message, the creators of advertising also support their arguments in favor of the promoted product or service by referencing scientific research. The perceived believability of such ads was examined by Beltramini and Evans (1984). Their research has shown that consumers are more likely to trust ads presenting results of scientific research, although executed within small research samples. Moreover, believability was not influenced by the information if the presented research was executed by an independent institution, or the advertiser himself, or by an expert. Quantitative data was perceived as more believable than qualitative data. The believability of information may not always declare the effectiveness of advertising. Information on social anti-smoking campaigns—about the harmful effects of cigarette smoke—is believed by respondents, but studies show that such campaigns are still ineffective (Calvert, Gallopel-Morvan, Sauneron, & Oullier, 2010). When an audience receives rational or truthful information, there still may be a negative response. Research by Mehta (2000) confirmed that the advertising is effective if the target audience likes it, and at the same time believes in it, finding benefit for current products and services. Trust can also be increased by partial elements of marketing communication such as the font type (Brunner-Sperdin, Ploner, & Schorn, 2014), the colors of ads (Lee & Rao, 2010), and also eye contact of the person on the cover, recently proven in the research of Aviva, Tal, and Wansink (2014).

Although trust is important in all areas of marketing communication, the key areas where trust is essential, is the banking sector. The trust of ordinary clients in banks depends on the concrete bank institution as well as on global aspects, such as world financial and economic crises. As in other areas, in banking there are efforts by researchers (see e.g., Cho, Huh, & Faber, 2014) to identify the essential factors that support or build the trust of bank clients. According to Delina and Svocáková (2013), there also exists a relationship between the trust to Slovak and to world banks. Conversely, the length of commercial relationships does not affect the level of trust. Experiences of clients are a valuable source of information that helps to build the credibility of an institution. This is one of the reasons why this type of information gets the attention of marketers. This trend is followed by attempts to assess the credibility of the institutions by an independent assessor. In Slovakia, there is an independent certification company *Research Center of Customer Views*. It works by evaluating the response of consumers and business partners in the company.

One of the ways to enhance the credibility of marketing communications is the use of a suitably chosen celebrity, whose function in marketing is to communicate as the referrer. This subject is discussed in the next section of this chapter.

Celebrities in Advertising and Their Credibility

Today, endorsements by celebrities are an established element in brand and product communication. The most famous definition of a celebrity endorser is a definition by McCracken (1989): a celebrity endorser is a person who ". . .enjoys a public fame and such a fame is used in a way that he/she appears in advertisements for a promoted product" (p. 310).

Various authors dealing with celebrity endorsements (e.g., Shimp, 2007) apply the premise that efficiency of this promotional form is influenced by two basic attributes: credibility and attractiveness. Subsequently, two models were created: the source credibility model and the attractiveness model. Both elements were understood as independent ones, with some measure competing of attitude variations toward the relationship between an endorser and its audience. The source credibility model or theory is the oldest attempt to explain how celebrity endorsement functions. The core postulate of the model is the dependency of successful communication on source quality. The source credibility model is based on research in social psychology.

The essence of credibility from the point of its role in the endorsement process is an assumption that recipients would rather accept a message that they can trust (Tellis, 2000). Credibility consists of two sub attributes: trustworthiness and expertise. Trustworthiness is linked to reliability, integrity, and trustfulness of a communicator, ". . . it is an ability of a source to transfer honest, objective utterances" (Tellis, 2000, p. 256). The endorser's trustworthiness depends to a great measure on how the audience decodes his/her motivation for brand promotion. A celebrity gains the trustworthiness of the audience via his/her life—primarily professional (on screen, in a sports game, etc.), but secondarily also a private one (or at least its aspects that are shown to the public via the media). Both such sides of life are interconnected and strongly influence one another. In general, celebrities taking on the role of endorsers must present themselves as the ones who do not manipulate the audience and are utterly objective and honest. The overall effectiveness of the celebrity endorsement process (Tellis, 2000) then lies in whether the recipients' trust toward a celebrity can overcome prejudice rooted in the fact that a celebrity endorser gets paid or they start to believe that he/she also uses the products/brands endorsed (Ohanian, 1991 and others).

A second credibility attribute is expertise or professionalism. The attribute is related to knowledge, experience, or the abilities of the endorser that have a link to the promoted brand. There is a capability of the source to communicate valid and unbiased statements. Shimp (2007) emphasizes that professionalism is not an absolute phenomenon, but rather depends on a subjective perceptive ability of the audience.

If the audience evaluates a celebrity endorser as a credible one, there is a high probability that the advertising message will be positively accepted. The process is labeled internalization—acquisition and acceptance of the communicated idea and its integration in the recipient's existing value system and his/her attitudes (Světlík, 2012). Internalization happens when the recipient acquires the presented attitude toward the promoted brand and accepts it as his/her own.

An important factor that is necessary to consider in relation to celebrity endorsers is the question of celebrity saturation. Some authors believe this factor to be the key (e.g., Um, 2008) for effective communication. We understand saturation as a level of celebrity involvement in an ad. At times—it is especially true with extremely popular celebrities—a celebrity can be involved in the promotion of too many brands. In such a case, we can talk of multiple endorsements and overexposure of his/her marketing and communication potential that leads to attenuation of the link to the promoted brand. Another form of multiple endorsements is the promotion of one brand/product by several celebrities (underexposure). The first case includes sharing one celebrity among more brands. This practice raises a question as to how this impacts audience perception of a brand and the endorser. Research has demonstrated that (Mowen & Brown as cited in Um, 2008; Tripp, Jensen, & Carlson, 1994) celebrity overexposure has negative consequences on the celebrities and commercials in which they appear on the one hand, but also negatively influence the assessment of the brand and shopping intention on the other hand. Another important factor is whether the endorsed brands/products appear in the same or similar product segment or they belong to completely different segments. Tripp et al. (1994) mention that endorsing more than three brands has a negative impact on celebrity endorsement credibility, which is true for more famous as well as for less famous celebrities. This has a negative impact on the endorser's credibility. The trustworthiness of a celebrity decreases together with his/her high involvement in a commercial, as recipients may link such an attempt with dishonesty not only to the promoted brand but also to his/her audience. Secondarily, negative qualities can be also transferred to the brand/product.

Creativity in Advertising and Trust

Creativity in advertising is one of the aspects that helps to attract the attention of the recipient (see research by Wilson, Baack, & Till, 2011) and to develop a positive attitude toward advertising, and secondary to the presented service, product or idea. Some sources, for example, Wells, Burnett, and Moriarty (1965), Sankaran (2013) state that ". . .creativity of advertising can be counter-productive and there may occur so called effect of the vampire creativity" (p. 2). In case advertising is not useful and applicable we cannot identify it as being creative. It can be original, but not creative. So, the assumption of "vampire creativity" is incorrect. Only "vampire originality" can be considered. Toward that argument it must be argued that it violates the current definition of creativity—if we follow the valid definitions (Rothenberg & Hausman, 1976; Szabo & Szabová, 2014; Szobiová, 2004; Wąsiński, 2005), so we could consider that creative ads are only those which are characterized simultaneously by *originality and utility* (suitability and application). Among respondents, the novelty and utility of creative advertising affected the attitude toward the advertised brand, while its utility affected the trust in the brand positively (see Sheinin, Varki, & Ashley, 2011).

Bellman, Robinson, Wooley, and Varan (2017) proved in their research that perceived creativity of advertising increased the probability of acceptance of the advertised message. Ang, Lee, and Leong (2008), have stated that new and meaningful advertising with which viewers can "connect" can produce higher data recovery and more favorable attitudes toward advertising compared with advertising that does not dispose these qualities. Similar results were reached by Reinartz and Saffert (2013) who realized their research in Germany. The research significantly proved that creative texts reached a higher level of consumer attention, leading to a more positive view of the brand.

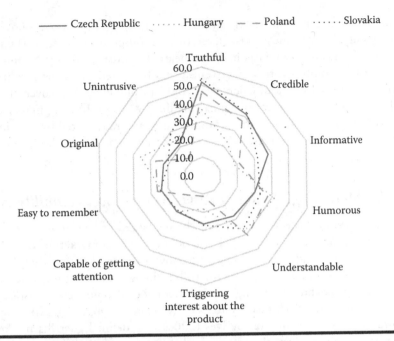

Figure 9.1 Customers' requirements for advertising in the V4 countries. (Authors' processing of the data according to Šimek, I., & Jakubéczyová, *Postoje verejenosti k reklame. Medzinárodná štúdia zo strednej Európy* [Public attitudes to advertising. International studies from Central Europe.] Taylor Nelson Sofres SK. Retrieved from http://www.tns-global.sk/docs/Postoj_k_reklame1.ppt.)

Consumer surveys carried out in 2003 within the V4 countries reported on advertising requirements (by Taylor Nelson Sofres SK, reported by Šimek & Jakubéczyová, 2003). Results proved that besides truthfulness, informative value, and the competence to attract attention, triggering interest is also necessary in order *to be* credible, *humorous,* and *original* (see Figure 9.1). Novelty and originality of the approach are therefore conditions for the ultimate success of advertising communication.

From all forms of marketing and communications (see e.g., Štrbová, 2012; Tomczyk, 2010), only some suggest creativity by its very nature. By using specific elements of the environment and by its essentially unrepeatable character (always use what the current environment offers and allows), ambient advertising fulfills the attribute of originality as the main principle that is fundamental for creativity. Ambient advertising communication uses nontraditional or nonstandard alternative media, so-called ad hoc, that is used only once, for certain purposes; for example, the advertising on public transportation buses, hairdresser salons, and many others (see e.g., Wojciechowski, 2014, 2016). This type of communication adequately addresses the target groups and it is not likely that recipients will become oversaturated by this communication form (Lehnert, Till, & Carlson, 2013) whereas it is unrepeatable. Jurca, Romonți-Maniu, and Zaharie (2013) compared the effectiveness of ads located in the nontraditional and unconventional medium (such as an ambient medium). Moreover, the perceived credibility of an unconventional medium was significantly higher compared with a traditional medium. Similar findings were reported by Maniu and Zaharie (2014). For these reasons, the next section is devoted to ambient marketing communication.

Guerilla Marketing, Ambient Advertising, and Credibility

Jay Conrad Levinson used the term and thesis of *guerilla marketing* as we know it now in his book published in 1984. He defined guerilla marketing as unusual, unconventional methods of marketing with a low budget and "going after the conventional goals of profits, sales and growth but doing it by using unconventional means, such as expanding offerings during gloomy economic days to inspire customers to increase the size of each purchase" (Levinson, 2003, p. 4). To compare, we can also present a definition by Ives (2004) who defined guerrilla marketing as "A broad range of advertising methods that strives to strike when people least expect it. Though publicity stunts have been turning heads for ever, mainstream marketers are increasingly turning to guerrilla tactics as consumers prove more difficult to reach with traditional advertising" (p. 23). Guerilla marketing belongs to unconventional forms of marketing, the aim of which is to address and attract potential clients and at the same time to maintain a low budget.

Guerilla marketing (see Figure 9.2) is based on unconventional ways of promoting products, services, and ideas. Therefore, it is considered to be a nonstandard form of marketing. It is a form of promotion that cannot be avoided by even the largest opponent of advertisements. According to Světlík (2012) and Szyszka (2013), these forms can affect everyone, everywhere (also in campaigns of cultural institutions [Walotek-Ściańska, 2015] or NGOs) and using, for example, word-of-mouth marketing, social media marketing, viral marketing, ambient marketing, and public relations (for more details, see Ďurková, 2011 or Gajdka, 2012).

Therefore, Hutter and Hoffmann (2011) present benefits that have their basis in a beneficial relationship of costs for promotion and resultant effects. To achieve this target, the marketing campaigns primarily focus on so-called "Surprise Effect" in their first phase, which they achieve by creating a new and unconventional idea (Frankova, 2011). This idea must be communicated to the target group. This phase is followed by the next phase, called the "Diffusion Effect" in which the aim is to stimulate consumers and/or media to spread the message further. Another effect, called the "Low Cost Effect" means that the increasing Surprise Effect is aggregated by several persons in the target group, which finally decreases relative costs.

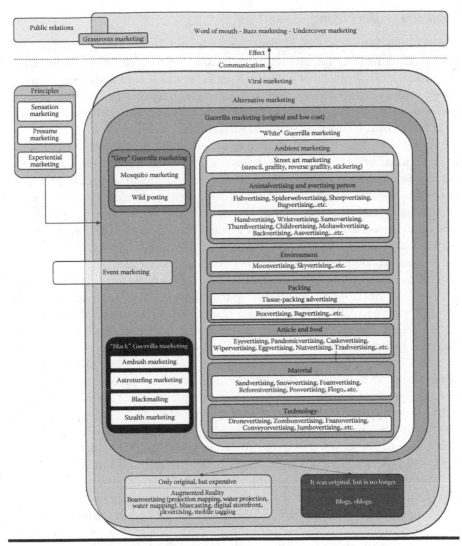

Figure 9.2 Relations of guerilla marketing forms by Wojciechowski. (From Wojciechowski, Ł. P., *Ambient Marketing + Case Studies in V4*, Towarzystwo Słowaków w Polsce, Kraków, Poland, 2016.)

Several types of guerilla marketing and tools of ambient marketing exist in different symbiotic correlations (Solík, 2014). Ambient marketing tools also point to the level of their similarity, which is expressed by the way of their placement, implementation, time characteristics, cost, material design, and technological support (Pavlů, 2009; Mago, 2015; Štrbová, 2012). Therefore, we can speak about a certain coherence thanks to which these techniques can be assigned to ambient forms.

Ambient Advertising

Ambient marketing is a term that is mainly and most often connected with nonstandard, imaginative, and innovative marketing. Despite the problem with its definition, it usually includes activities from fields which have nothing in common (Wilczek & Fertak, 2004). According to Luxton and Drummond (2000), the term *ambient* was used in the context of marketing communication in 1996 for the first time by the British company Concord Advertising.

At the beginning, marketers started to use places that had never been considered as places for advertising, such as floors, fuel dispenser nozzles, fountains, rubbish bins, benches (*street furniture*), interior doors of public toilets, shopping carts, large surfaces of buildings, and so on. However, the selected localization should be associated with a message of the advertisement, so use of an untraditional place was not purposeless. It assigns them specific new meanings in a significantly unconventional way.

Hatalska (2002) writes: "ambient media . . . include all non-standard activities performed by classical media and other communication channels" (p. 7). Jurca and Plăiaș (2013) characterize this form as follows: "ambient advertising refers to a creative form of out-of-home advertising that conveys direct and contextual messages by using and often altering existing elements of the environment in a way that surprises the target audience" (p. 1769). Advertisements are placed in unusual areas directly in the social environment of a target group, where they live, work, and play (Krautsack, 2008). According to Barnes (1999), a characteristic feature of ambient advertisement is to surprise a consumer by confrontation with incongruent suggestions on an unusual background. The only certain fact is that ambient marketing is based on unconventional, new, and surprising ideas and uses unconventional placement (Derbaix & Vanhamme, 2000; Meyer & Niepel, 1994; Pavel & Cătoiu, 2009; Lee & Dacko, 2011; Luxton & Drummond, 2000; Shankar & Horton, 1999). Surprise is a result of the difference between perception and expectation.

Moreover, we also observe a significant decrease of consumers' confidence toward advertising a communication message. According to a survey conducted by Nielsen (2013) in 56 selected markets around the world, the greatest potential in sales uses *buzz marketing* as a form of advertising (Hughes, 2008). Almost 84% of respondents decided that the most important sources of information are recommendations and references from other consumers when deciding on a purchase.

Ambient advertising is a certain form of creative thinking (Csikszentmihalyi, 1996; Fichnová, 2013). It needs creativity, and so it unintentionally becomes a pioneer for new trends (Ogonowska, 2014). At the same time, the life of a specific ambient advertisement does not last forever. This kind of advertisement with a specific communication message can be effective only for a short period.

However, ambient marketing has a much greater potential to affect a possible recipient than any other form of advertisement not only because it is different, but mainly because it directly addresses the recipient. This is another advantage of this form of marketing.

Generations X and Y and Their Specific Characteristics

When specifying certain common characteristics among age cohorts, we can rely on the findings of psychologists. As some research suggests, in every new generation there occur certain specifics related to the era, in which the individual grows and his personality is therefore undoubtedly influenced by wide social, technological, cultural, or political aspects. Some refer to this trend as a secular trend (e.g., Mingroni, 2004; Rösler, 1990), or the acceleration of mental development. Others refer to specific generations (e.g., Jandourek, 2001). The term generation refers to the group in which individuals of the same or similar age are ideologically and spiritually connected and their so-called "horizon of vision" is determined by similar historical destiny (Mannheim, 1952). Social, political, cultural as well as changes in the development of science and technology are specific for different periods and provide a framework in which individuals of a given generation develop and form.

The evident advancement of technology, digitalization, and advanced penetration of computers, the Internet, and other conveniences of the present world are hence becoming one of the factors forming individuals and society. McLuhan's (1991, p. xxi) statement that: "We shape our tools, and thereafter our tools shape us" fittingly describes this state. A McLuhan follower, media analyst Postman (1999), advocates the thesis saying that every new kind of media in history (script, letterpress, telegraph, photography, radio, television) has changed the structure of discourse in society. However, as many authors have insinuated (e.g., Gołuchowski & Losa-Jonczyk, 2013; Mistrík, 2004; Vybíral, 2005), technologies do not only characterize culture, but they also influence and shape communities as well as individuals and their behavior.

Prensky (2001) described youths as "digital natives" and those, who were born and raised in the predigital era as "digital immigrants." Members of the "digital natives" generation have their own ways of learning as well as preferred cognitive styles (there exist objections against this approach, e.g., Bennett & Maton, 2010). Whether we will refer to this generation with Prenskys' term "digital natives" (2001) or rather spread the label "Net-generation" (Leung, 2004; Tapscott, 2009) or "Millenials" (Oblinger & Oblinger, 2005), we do not indicate the demographic demarcation, but more likely the complexity of the demographic cohort, its values, life experiences, and life style.

These terms designate the first generation that grew up surrounded by technology such as personal computers, video games, and the Internet. The first to define specifics for each generation was Howe and Strauss (1997) in the publication *The Fourth Turning* in which Generation Ys are characterized as Millennials, and according to these authors it is bordered by the years 1982–2004, whereas Leung (2004) present this group as young people born between 1977 and 1997. The exact time demarcation of this generation has not been unified thus far and opinions of various authors differ—from the beginning in 1977 (Bakewell & Mitchell, 2003), to 1982 (McCrindle, 2006). Similarly, the upper borderline varies from 1990 (Horváthová, 2012) to 2004 (Howe & Strauss, 1997). The same situation occurs when defining Generation X. For the purposes of this chapter, we will follow the definition of generations given by Howe and Strauss (1997). Considering the fact that in every country there can be various characteristics of generations regarding the current socioeconomic, political, and technological changes, this is not an ideal concept. However, it also helps to identify the generation in the wider geopolitical context and in view of globalization trends in culture, business, finance, or marketing communications.

The questions of Generation Y and its specifics are often being presented, whereas there exists little empirical research. Researchers such as Ball, Beard, and Newland (2007) support the focus of this generation on new technologies, which was manifested in their research as a decline in an interest in the classical types of books along with a 146% increase in an interest in using electronic full textbooks and journals. Similar findings were introduced by Beard and Dale (2008) and Gálik (2010). In view of these changes and the specifics of Generation Y compared with Generation X we can also consider their specific preferences with respect to the media (this area was examined in the Slovak population, e.g., by Rábeková, 2008). Various media preferences among different generations will likely impact the preferred type of marketing communication. These preferences may not be sufficiently predictable, as demonstrated in our previous research (Fichnová & Wojciechowski, 2016) where the expected preference of digital communication has proved to be perceived as worse than ambient communication.

Like the preference of the media, preferred types of patterns (idols) may be differentiated per generation segments (Vrabec & Petranová, 2013), which can then be important information when deciding which famous person to engage in the content of marketing communications.

Research

Research questions (RQs): Based on the abovementioned facts, we conceived several research questions, and we divided them into several groups as follows:

1. Trust in Publicly Known Figures
   RQ1: Which publicly known figures in Slovakia are credible for the representatives of Generation X?
   RQ2: Which publicly known figures in Slovakia are credible for the representatives of Generation Y?
   RQ3: What reasons have led respondents to their choice?
2. Creativity of Ads and Their Credibility
   RQ4: Is there relationship between the degree of creative realization of advertising and its perceived credibility?

Methodology

Methods of Data Collection and Procedure

For data collection, we prepared our own combined type of questionnaire. Questions were aimed to identify the name of the public figure evaluated by the respondent as credible and the reasons that led the respondent to his choice. The questionnaire also contained the questions in which the interviewed people valued various types of ads (print ads used in the daily press and magazines, outdoor ads/billboards/ambient ads, audiovisual spots, videos on social networks). Every ad was valued on the 5-point Likert type scale. These ads were designed to present different levels of creativity (creativity was set by experts according to the modified version of the evaluation scale of ads by Smith, MacKenzie, Yang, Buchholz, & Darley, 2017). The questionnaire was presented to the respondents in the classic print version. Another group of respondents was examined with an electronic version of the questionnaire, using a panel brought by Survio service (http://www.survio.com/sk/).

Methods of Data Analysis

The data obtained from the questionnaires were analyzed using several procedures. The open-type questions were processed by content analysis—within the modified version according to Scherer (2004) as the quantified content analysis. To analyze the frequency of important words, in some items of the questionnaire we used a tool that is available online https://www.online-utility.org/ (Adamovic, 2006). To evaluate the responses, we used descriptive statistics as well as methods of inductive statistics: Q correlations, t tests (double-paired and unpaired tests) with the Bonferroni correction and chi-square.

Research Sample

The sample consisted of a total of 177 respondents aged 18–54 (average age 33.33 years) with a balanced representation of gender, regions of Slovakia, and representation of respondents with different levels of educational attainment, while the percentage copied the demographic variables in this factor in Slovakia (data from the Slovak Statistics Bureau). Regarding research objectives, the research sample was divided into two cohorts. The first one consisted of respondents belonging to Generation X (born between the years 1961–1981). Regarding the different opinions to limit the age scale of this generation we used the criteria of How and Strauss (1997), who obviously came first with the generation concept. The total number of respondents was 75, the average age was 45.31 (sd = 6.81; mod = 39; med = 43). We applied the same procedure for Generation Y (born between the years 1982–2004). The total number of respondents was 102, the average age was 27.53 (sd = 2.82; mod = 20; med = 21).

Results and Discussion

Trust of Generation X in Public Figures

A summary of the responses is provided in Table 9.1. Only 14 respondents (18.7%) did not have or did not mention any public figure that they could trust. A certain skepticism is reflected in the answers, *I do not know whether any public figure can be trusted, I do not trust anybody,* or *I trust only myself.* Two respondents (2.6%) reported more than one answer. Respondents representing Generation X in Slovakia reported up to 24% of the responses were politicians.

The most reported was Slovak President Andrej Kiska (38.9% all in the politician category). Respondents of Generation X mentioned the name of Russian President Vladimir Putin three times. Actors and spiritual persons were reported by respondents in the same frequency (11 times in both cases, that is, 14% of the total number of Generation X respondents). The most frequently reported was the actor Milan Lasica (5), the most reported spiritual persons were Pope Francis (3) and Dalai Lama (3).

Singers, presenters, or scientists were reported four times each. The Generation X respondents did not name any writers, but instead a recently deceased visual artist, architect, designer, and university teacher of Czech origin was named. He was well known as the creator of glass objects, furniture, and the main administrator of the Prague castle during the reign of Václav Havel. From these results, it is difficult to determine the most suitable candidate to take advantage of well-known personalities (public figures) in marketing communications in the role of an endorser (recommender).

Frequent use of sportsmen (especially Peter Sagan) only reached a low percentage of representatives from Generation X. On the other hand, there was a high potential for political communications, as politicians were most frequently reported by this cohort as credible personalities. Regarding the low number of preferred figures, the generalization of the obtained results was not possible. Generation X is not homogenous in these preferences, but some trends, as we presented above, are obvious.

Trust of Generation Y in Public Figures

Based on the quantitative analysis (when the unit of analysis was a specific response to an item in the questionnaire), we found that the Generation Y sample could be divided into three smaller groups that were not equally numerous, but reflected their views on the credibility of public figures. A summary of the results is presented in Table 9.2.

In the first group of "distrusting" respondents (36 answers in total, relative quantity—further "RQ" 35.3%) there were either no answers to the question (6.86%) or a distrust in public figures (16.67%). Part of the respondents (5.9%) stated

Table 9.1 Public Figures Trusted by Generation X Respondents in Slovakia

Gen X		Examples of Answers	Absolute Frequency[a]	Relative Frequency
One public figure		59 respondents	59	78.7%
Two answers		2 respondents	2	2.6%
No figure or no answer		14: no answer (1), no figure (3), do not know such figure (3), trust in nobody (1), do not know (2), do not know if any public figure is credible (1), do not trust anybody (1), family (1), trust in oneself (1)	14	18.7%

The chi-square value is 72.24. The p value is <0.001. The result is significant at p = ≤0.01.

Singer, Musician				
	Foreign	Karel Gott (1), Madonna (1)	2	5.3%
	Slovak	Marika Gombitová (1), Zuzana Smatanová (1)	2	
TV Presenter, Editor				
	Foreign		0	5.3%
	Slovak	Adriana Kmotríková (1), Vilo Rozboril (2), Michal Hudák (1)	4	
Actor/Actress				
	Foreign	Antonio Banderas (1), Jackie Chan (1), Code de Pablo (1)	3	14.7%
	Slovak	Milan Lasica (5), *Ladislav Chudík* (1), *Jozef Króner* (1), Zdena Studénková (1)	8	
Writer, Press Editor			0	
Spiritual				
	Foreign	*Mother Teresa* (1), Pope Francis (3), Dalai Lama (3)	7	14.7%
	Slovak	Archbishop Bezák (1), *Anton Strholec* (1)	2	
	Jesus	*2*	2	

(Continued)

Table 9.1 Public Figures Trusted by Generation X Respondents in Slovakia (*Continued*)

Gen X		Examples of Answers	Absolute Frequency[a]	Relative Frequency
Sportsman				
	Foreign		0	9.3%
	Slovak	Peter Sagan (3), Marián Hossa (1), Zdeno Chára (1), Adam Žampa (1), Dominika Cibulková (1)	7	
Politician				
	Foreign	Vladimir Putin (3)	3	24%
	Slovak	Andrej Kiska (President of SR) (7), Milan Kňažko[b](2), Róbert Fico (1), Ján Kotleba[c] (1), Daniel Lipšic (1), Július Binder[d] (1), *Milan Rastislav Štefánik* (1), Marek Krajčí (1)	15	
Others		Sergei Nikolajewitsch Lasarew (1) (parapsychologist), Monika (1), Tina Zlatoš Turanová (1)	3	4%
Visual artist		*Bořek Šípek* (1)	1	1.3%
Scientist		Prof. MUDr. Vladimír Krčméry, Dr.Sc. – medical doctor, scientist (1), *Albert Einstein* (2), *Nikola Tesla* (1)	4	5.3%

Note: Italics are used to mark public figures who are already deceased.

[a] Note: Some respondents reported more than one answer, despite the instructions. For this reason, the total sum is not equal with the total number of respondents. The sum of percent—relative size is not 100 for the same reason. We calculated the percentage as the number of relevant responses to the number of respondents.

[b] M. Kňažko was originally known as an actor. At the time of the Velvet Revolution (1989), he engaged in the movement Public against Violence, which he co-founded. He also co-founded the political parties *HZDS* (1991) and *SDKÚ* (2000). In 1998–2002, he was Slovak Minister of Culture; in 2014, he was a candidate for Slovak President.

[c] On the Slovak political scene are two names with the same surname: Kotleba (Ján and Marián).

[d] J. Binder was classified as a politician regarding his political engagement, a member of the National Council in 1998–2002, a candidate for mayor of Bratislava.

Table 9.2 Public Figures Trusted by Generation Y Respondents in Slovakia

Gen Y		Examples of Answers	Absolute Quantity (Total)[a]	Relative Quantity
One public figure		61 respondents	**61**	**59.8%**
Two or more answers		5 respondents	**5**	**4.9%**
No figure or no answer		36: No answer (7), none—such figure does not exist (17), cannot judge, do not know personally, not possible to find out on TV (6), do not remind (1), people I trust in are not public figures (1), it is not possible to trust in public figures (1), all are liars, I cannot trust any, I think everybody pretends (2), I do not trust in anybody (1)	**36**	**35.3%**

The chi-square value is 46.294. The p value is <0.001. p = ≤0.01

Singer, Musician				
	Foreign	Adel (1), Armstrong Billie Joe (1), Bono (1), Demi Lovato (1), Jennifer Lopez (1), Scherzinger Nicol (1), Valentina Lisitsa (1)	7	10.8%
	Slovak	Dara Rolins (2), Kristína Peláková (1), Tina (1)	4	
TV Presenter, Editor				
	Foreign		0	19.6%
	Slovak	Adela Banášová (11), Matej Cifra –Sajfa (2), Patrik Švajda (1), Vilo Rozboril (2), Zlatica Puškárová (4)	20	
Actor/Actress				
	Foreign	Chuck Norris (1), Angelina Jolie (3), Leonardo Di Caprio (1)	5	5.9%
	Slovak	Kristína Farkašová (1)	1	

Gen Y		Examples of Answers	Absolute Quantity (Total)[a]	Relative Quantity
Writer, Press Editor				
	Slovak	Boris Filan (1), Renáta Klačanská (1), Evita Urbaníková (1)	3	2.9%
Spiritual				
	Jesus	2	2	4.9%
	Foreign	Pope Francis (3)	3	
Sportsman				
	Foreign	David Beckham (1), Hosszú Katinka (1), Ussain Bolt (1)	3	9.8%
	Slovak	Dominika Cibulková (1), Radoslav Židek (1), Miroslav Šatan (1), Matej Tóth (1), Peter Sagan (1), Marek Viedenský (1), Vlado Zlatoš (1)	7	
Politician				
	Foreign	Barack Obama (2), Queen Elizabeth (1), Vladimir Putin (1)	4	14.7%
	Slovak	Andrej Kiska (president of SR) (5), Boris Kollár (1), Marián Kotleba (2), Igor Matovič (2), Richard Sulík (1)	11	
Others		Petra Straková (1), Gary Vaynerchuk (1), *Maria Theresa* (1)	3	2.9%
Visual artist			0	0%
Scientist			0	0%

[a] Note: Some respondents give more than one answer, despite the instructions; for this reason, the total sum is not equal to the total number of respondents.

that they trusted only those people who they personally knew, which indicates the individual experience and that two-way communication with some sort of relationship is important for building trust. Other respondents expressed extremely strong distrust (even disappointment) with answers such as *All are liars, I do not trust anybody, I do not have confidence in people that are public figures,* and *All are pretending.* The question is whether this results in some personal disappointments and its generalizations, transposed opinion, the general incredulity, or other factors.

The second group could be called "standard trusting" (59.8%), where the respondents claimed only one public figure.

This subgroup was larger than the first "distrusting" subgroup. This difference was statistically significant. The representatives of Generation Y in Slovakia trust mostly the Slovak presenters—editors (19.6%). The presenter Adela Banášová received the highest number of votes (10.78%), followed by Zlatica Puškárová with 4 votes. The secondmost trusted subgroup were politicians (14.7%). The most trustful among the politicians was the president of Slovakia, Andrej Kiska (5); Parliament members Igor Matovič (2) and Marián Kotleba (2); and former U.S. President Barack Obama (2). The next most numerous group of public figures trusted by respondents were musicians (11) and sportsmen (10), whereby musicians were more often foreign and sportsmen more often of Slovak origin.

The last, the least numerous group, consisted of respondents indicating more than one trustworthy public figure. Answers included two to three public figures from the same public sphere (e.g., politics), or from different spheres. It is worth noting that in only three cases, did Generation Y respondents mention the name of a writer. They did not mention any scientists or visual artists.

Comparison of Trust of Generations X and Y in Public Figures

According to analysis of data provided in Tables 9.1 and 9.2 and their comparison (Table 9.3), there are obvious elementary differences in which public figures are trusted by Generations X and Y in Slovakia. Generation X respondents are willing to trust politicians (almost one-quarter of them), while the Generation Y respondents trust TV presenters and editors (19.6%) and politicians (14.7%) as well. Another difference appeared in the frequency of voting for spiritual personalities. Generation X mentioned them more often than Generation Y.

Table 9.3   Trust in Public Figures by Generations X and Y According to Their Own Statements

	State at Least One Public Figure	Do Not Trust Public Figures	Marginal Row Totals
Gen X Group	61 (53.81) [0.96]	14 (21.19) [2.44]	75
Gen Y Group	66 (73.19) [0.71]	36 (28.81) [1.79]	102
Marginal Column Totals	127	50	177 (Grand Total)

Note: The chi-square statistic is 5.8954. The *p* value is .015181. This result is significant at *p* <.05.

The numerous group for Generation Y presented singers and musicians (10.8%), but in Generation X there were only 5.3% respondents. Generation Y respondents did not mention any scientists or artists, but among the Generation X respondents such answers appeared, however in small quantities (5.3% and 1.3%). In total, Generation Y respondents were less willing to trust public figures than Generation X respondents. Up to 35.3% representatives of Generation Y did not trust public figures because no suitable figure exists for them. The same answer was given by only 18.7% Generation X respondents. We subject the results of quantified content analysis to statistic comparison and based on this comparison we can state that the surveyed groups of respondents significantly differentiate in numbers of how many representatives of individual groups do not trust famous personalities—representatives of Generation Y prove to be less trusting. This finding can be attributed to their focus on personalities from the entertainment industry (musicians from abroad, or speakers of entertainment programs) who are usually mentioned by the media only because of scandals.

Reasons for Trust in Selected Public Figures Stated by Both Generations X and Y

Both generations identically stated those reasons that are mainly focused on either real or dedicated attributes of famous personalities. Generation X mentioned attributes evaluating personality dimensions: honesty, integrity, fairness, incorruptness, modesty, humbleness, justness. Generation Y also mentioned attributes of a personality or his/her features, but they were more of the emotional and social type: likeability, kindness, spontaneity, pleasantness, friendliness, humanity, and tactfulness. They also mentioned wisdom, sensibleness, erudition, and education. Besides the attributes, both generations also mentioned activities done by celebrities for which the generations considered them to be credible. Generation X mentioned: he/she makes surveys, fulfills promises, helps, acts. Generation Y most often mentioned that a celebrity is credible because he/she is involved in charity activities, helps people and children, takes care of other people, makes things with passion, are willing, and had never lied. This generation included the fact that such personality "records his/her life and publishes his/her videos" among relevant activities. Both generations also derived credibility from the status of a famous personality; for example, they state that a famous person is a representative of something, or a head of something, he/she knows something, manages something, he/she has a position, age, he/she is a good sportsman, writer, he/she has achieved something, and is successful. Similarly, both generations relied on their internal feelings—Generation X stated: *I have such feeling of him/her, He/she acts on me like that,* and *Nice facial features.* Generation Y stated: *I like him/her, He/she does not mislead, He/she acts on me like that, He/she performs in public like that, His/her voice is interesting,* and *His/her opinions are good.* Generation X also stated: *He/she can motivate me.* Besides the categories mentioned above, Generation Y also had another category which could be called a nonscandalous way of life. They considered a famous personality to be credible if he/she is decent, without any scandals or with a minimum number of scandals, or without a big scandal.

Creativity of Advertising—Ads and Their Trust

The research was also designed to explain possible relationships between the measurement of creative advertising and their perceived credibility. Rate of creativity in ads was evaluated through a modified scoring scale by Smith et al. (2007). We used criteria as follows: originality, flexibility, synthesis, elaboration, artistic value, ad-to-consumer relevance, and brand-to-consumer relevance. Every criterion was evaluated on a scale of 1 to 6. The final score showed the creativity level of each of the eight ads. The mentioned authors present preliminary norms, so it was possible to objectively identify which ads could be considered creative and which ads were below the average. For further analysis, we chose two ads. Those were the ones that were characterized by extreme values for the dimensions: highly creative–low creative. For highly creative advertising, we chose C (ambient advertising using fountains) and as a low creative ad, we chose G (billboards). Next, based on a questionnaire (described in the "Methods" section) we investigated the level of perceived credibility of ads by the two groups (the average assumed credibility of all ads, particularly the average values as well as other statistical indicators). Contingency Tables 9.4 and 9.5 present the response rate of the respondents—in assessing the credibility of creative and uncreative advertising. The respondents were not familiar with the categorization of ads, so their evaluations were based solely on their own judgment.

Results confirmed a significant relationship between the level of creativity of advertising and its credibility in both groups. The more interesting the advertising for its creativity, the larger the tendency of spreading the message and trust of the advertising. In this concrete case, creative advertising was identified as the ambient advertising sample.

It is obvious that this kind of advertising is more effective to recipients in gaining their trust. By analyzing the results more precisely, we can conclude that despite some expected greater conservatism of the older generation, our results suggest that highly creative ambient advertising was considered as one the most trusted advertising mediums by Generation X.

On the contrary, Generation Y representatives evaluated it as "lukewarm" (2.63), although its credibility was ranked in third place out of eight ads. The low value isn't a reflection of untrustworthiness of that ad (or ads of this type), but rather a manifestation of the critical approach to advertising and marketing communication by all representatives of the younger generation in general. They were also more reserved in their choice of trustworthy well-known personalities, while several respondents answered that they didn't trust media-covered personalities at all. Nevertheless, the best rankings were given to media-famous domestic YouTubers.

Table 9.4 Contingent Table of the Relationship between Creativity of Advertising and the Perceived Credibility of Advertising by Generation X

Generation X	Minimal Trust	Low Trust	Average Trust	High Trust	Maximal Trust	n_j
Creative advertising	1 (4.50) [2.72]	4 (8.00) [2.00]	51 (43.50) [1.29]	16 (14.50) [0.16]	3 (4.50) [0.50]	75
Noncreative advertising	8 (4.50) [2.72]	12 (8.00) [2.00]	36 (43.50) [1.29]	13 (14.50) [0.16]	6 (4.50) [0.50]	75
n_i	9	16	87	29	9	150

Note: The chi-square statistic is 13.341. The *p* value is .009724. The result is significant at *p* <.01.

Table 9.5 Contingent Table of the Relationship between Creativity of Advertising and the Perceived Credibility of Advertising by Generation Y

Generation X	Minimal Trust	Low Trust	Average Trust	High Trust	Maximal Trust	n_j
Creative advertising	14 (20.50) [2.06]	30 (30.00) [0.00]	41 (31.50) [2.87]	11 (15.50) [1.31]	5 (3.50) [0.64]	101
Noncreative advertising	27 (20.50) [2.06]	30 (30.00) [0.00]	22 (31.50) [2.87]	20 (15.50) [1.31]	2 (3.50) [0.64]	101
n_i	41	60	63	31	7	202

Note: The chi-square statistic is 13.7507. The *p* value is .008135. The result is significant at *p* <.01.

Based on these paradox findings we can assume, that well-known personality or celebrity endorsers in combination with creative execution of ads are factors with a positive influence on the credibility of advertising.

Limitations of Research

First, basic limitation lies in the sample size included in research. For financial and organizational reasons, it was not possible to include a greater number of respondents and respondents from other countries. At the same time, classification into cohorts by generation brings pitfalls, relating to the fact that in different countries there are specific conditions of individual development based on cultural, political, economic, and other conditions. Therefore, the age definition of generations can be different in various countries. Also, the specifics of these factors create unique and nontransferable modification of specific symptoms with representatives of generations.

Conclusion

We identified the differences in preferences of two groups—Generation X and a younger Generation Y. Although results of the research have some limitations (particularly in the size of sample), we believe we identified at least some trends. Research was triangulated as we connected qualitative and quantitative techniques. We discovered the following:

- Generation X respondents often trust politicians, respondents of Generation Y trust Slovak guests—politicians together with TV news personas were only in second place.
- In general, Generation Y respondents can believe well-known personalities less than Generation X respondents.
- Both generations have similar reasons for trusting well-known personalities.
- One more category connected with perception of well-known personalities by Generation Y needs to be considered, and it is "nonscandalousness." A trustworthy well-known personality is someone, who, besides having attributes as being "polite," has "no scandals" or "minimum scandals," or "no past scandals."
- Attributes of celebrity endorsers are strongly connected with their credibility as perceived by the target group. We should also mention the attractiveness of a celebrity is another important component of their presence in advertising. There are three subcategories that can be distinguished within the attractiveness dimension: physical attractiveness, respect, and similarity. As pointed out in previous research (Mikuláš & Světlík, 2016), there is another aspect of successful communication using celebrity endorsement, which is an

execution of the ad itself, or the way a celebrity is acting for the product or brand.

■ Generation Y respondents were more critical in the evaluation of ads than respondents of Generation X.

■ Generation X respondents perceived ambient advertising as the best; Generation Y respondents preferred YouTube videos.

■ Both generations perceive billboards as the least trustful form of advertising.

■ Trust is significantly connected with creativity of advertising.

When building trust, it is appropriate to consider the conceptualization of trust by target groups and integrate it into the marketing message; for example, the observed absence of scandals for building trust among Generation Y, or emphasis on politeness with Generation X. When planning a media selection, marketers should prefer different communication media than traditional billboards. Besides the use of videos, we can point to ambient advertising, having the potential to be significantly supportive in building trust thanks to its creativity. In the future, it would be helpful to extend the implementation of research in other countries to identify specifics of generations influenced by current geopolitical and cultural situations as well as to give recommendations for improved communication on a global scale. Inspiring challenge for future research will lie in the focus on the generation older than Generation X and on the generation younger than Generation Y (Generation Z). Even after indicated attempts, the topic of trust and credibility will not be exhausted and we believe that it will address more and more attention of experts in marketing communications and other fields as well.

References

Adamovic, M. (2006). *Online utility—Free online software utilities*. [Online software]. Retrieved from https://www.online-utility.org/.

Ang, S. H., Lee, Y. H., & Leong, S. M. (2008). The ad creativity cube: conceptualization & initial validation. In J. J. Sierra, R. S. Heiser, & I. M. Torres (Eds.), *Creativity via cartoon spokespeople in print ads: Capitalizing on the distinctiveness effect*. Retrieved from file:///C:/Users/User/Downloads/JA.pdf.

Aviva, M., Tal, A., & Wansink, B. (2014). Eyes in the aisles: Why is Cap'n crunch looking down at my child? *Environment & Behavior, 47*(7), 715–733. doi:10.1177/0013916514528793

Bakewell, C., & Mitchell, V.-W. (2003). Generation Y female consumer decision-making styles. *International Journal of Retail & Distribution Management, 31*(2), 95–106.

Ball, D., Beard, J., & Newland, B., (2007). E-books & virtual learning environments: Responses to transformational technology. *Acquisitions Librarian, 19*(3–4), 165–182.

Barnes, J. (1999). Creating a difference with ambient media. *Admap, 34*, 46–49.

Beard, J., & Dale, P. (2008). Redesigning services for the Net-Gen & beyond: A holistic review of pedagogy, resource and learning space. *New Review of Academic Librarianship, 14*(1–2), 99–114.

Bellman, S., Robinson, J. A., Wooley, B., & Varan, D. (2017). The effects of social TV on television advertising effectiveness. *Journal of Marketing Communications, 23*(1), 73–91. doi:10.1080/13527266.2014.921637

Beltramini, R. F., & Evans, K. R. (1984). Perceived believability of research results information in advertising. *Journal of Advertising, 14*(3), 18–31. doi:10.1080/00913367.1985.10672953

Bennett, S., & Maton, K. (2010). Beyond the "digital natives" debate: Towards a more nuanced understanding of student's technology experiences. *Journal of Computer Assisted Learning, 26*(5), 321–331.

Brunner-Sperdin, A., Ploner, J., & Schorn, R. (2014). The influence of color, shape, and font formatting on consumers' perception of online drugstores. In J. Cotte & S. Wood (Eds.), *Advances in consumer research* (Vol. 42, pp. 357–360). Duluth, MN: Association for Consumer Research.

Calvert, G., Gallopel-Morvan, K., Sauneron, S., & Oullier, O. (2010). Dans le cerveau du fumeur: Neurosciences et prévention du tabagisme [In the smoker's brain: Neuroscience and tobacco prevention]. In O. Oullier & S. Sauneron (Eds.), *Nouvelles approches de la prévention en santé publique* [New approaches to public health prevention]. (pp. 86–107). Paris, France: Direction de l'information légale et administrative.

Cho, S., Huh, J., & Faber, R. J. (2014). The influence of sender trust and advertiser trust on multistage effects of viral advertising. *Journal of Advertising, 43*(1), 100–114. doi:10.1080/00913367.2013.811707

Cint. (2017). Survey [Online insights exchange platform]. Retrieved from https://www .cint .com/

Csikszentmihalyi, M. (1996). *Creativity: Flow and the psychology of discovery and invention.* New York, NY: Harper Collins Publishers.

Delina, R., & Svocáková, S. (2013). Dôvera segmentu mladých ľudí v bankový sektor SR [Trust of youth in Slovak banking sector]. *Scientific Papers of the University of Pardubice. Series D, Faculty of Economics and Administration, 27*(2), 34–46.

Derbaix, C., & Vanhamme, J. (2000, May 23–26). The "you know what?" syndrome—how to use surprise for gaining success. In B. Wierenga, A. Smidts, & G. Antonides (Eds.), *Proceedings of the 29th Conference of the European Marketing Academy.* Rotterdam: Erasmus University Rotterdam.

Diehl, S., Mueller, B., & Terlutter, R. (2007). Skepticism toward pharmaceutical advertising in the U.S. and Germany. *Cross-Cultural Buyer Behavior Advances in International Marketing, 18*, 31–60. doi:10.1016/S1474-7979(06)18002-3

Doney, P. M., Cannon, J. P., & Mullen, M. R. (1998). Understanding the influence of national culture on the development of trust. *Academy of Management Review, 23*(3), 601–620, doi:10.5465/AMR.1998.926629

Ďurková, K. (2011). Public relations (PR). In D. Petranová & L. Čábyová (Eds.), *Media relations II* (pp. 8–36). Trnava: Univerzita sv. Cyrila a Metoda v Trnave.

Fichnová, K. (2013). *Psychology of creativity for marketing communication.* Noailles, France: Association Amitié Franco-Slovaque.

Fichnová, K., & Wojciechowski, Ł. (2016, December 1). *Digital and conventional form of ambient marketing of cultural and educational institutions and their subjective perception of youth.* International Scientific Conference "Religion and culture in the digital media," Kraków.

Franková, E. (2011). *Kreativita a inovace v organizaci* [Creativity and innovation in the organization], Praha, Slovakia: Grada.

Gajdka, K. (2012). *Rzecznik prasowy w otoczeniu mediów. Teoria i praktyka* [A spokesman in the surrounding media. Theory and practice]. Kraków, Poland: Universitas.

Gálik, S. (2010). Virtualizácia reality v kontexte elektronických médií [Virtualization reality in the context of electronic media]. In S. Magál, D. Petranová, & M. Solík (Eds.), *K problémom mediálnej komunikácie I.: aktuálne otázky mediálnej kultúry: komunikačný diskurz* [*The issues of media communication I: Current issues of media culture: Discourse of communication*] (pp. 327–333). Trnava, Slovakia: Univerzita sv. Cyrila a Metoda.

Garbarino, E., & Johnson, M. S. (1999). The different roles of satisfaction, trust, and commitment in customer relationships. *Journal of Marketing, 63*(2), 70–87. doi:10.2307/1251946

Gołuchowski, J., & Losa-Jonczyk A. (2013). Wykorzystanie nowych mediów w promocji idei społecznej odpowiedzialności uczelni. [Using new media for the promotion of university social responsibility]. *Studia Ekonomiczne, 13*(157), 67–78.

Haller, T. F. (1974). What students think of advertising. *Journal of Advertising Research, 14*(1), 33–38.

Hatalska, N. (2002). Niestandardowe formy promocji [Non-standard forms of promotion]. *Marketing i Rynek, 11*, 7-12. Retrieved from https://goo.gl/Gs1fuO.

Horváthová, M. (2012). *Marketing tretieho tisícročia: zelený marketing* [Marketing of third millennium—Green marketing]. Retrieved from https://goo.gl/rk8z6R.

Howe, N., & Strauss, W. (1997). *The fourth turning. An American philosophy.* New York, NY: Broadway Book.

Hughes, M. (2008). *Buzzmarketing: Get people to talk about your stuff.* New York, NY: Portfolio Penguin Group.

Hutter, K., & Hoffmann, S. (2011). Guerilla marketing: The nature of the concept and propositions for further research. *Asian Journal of Marketing, 5*(2), 39–54. doi: 10.3923/ajm.2011.39.54

IAB Europe Report. (2016). *IAB Europe report: AdEx Benchmark 2015 – the definitive guide to Europe's online advertising market.* Brussels: IAB Europe. Retrieved from https://goo.gl/jzvkSO.

Ives, N. (2004, June 24). Guerrilla campaigns are going to extremes, but will the message stick? *New York Times*, p. C.6.

Jandourek, J. (2001). *Sociologický slovník* [Sociological dictionary] (1st ed.). Praha, Slovakia: Portál.

Jurca, M. A., & Plăiaş, I. (2013). *Schema congruity—A basis for evaluating ambient advertising effectiveness.* Retrieved from https://goo.gl/PHAQ5M.

Jurca, M. A., Romonţi-Maniu, A. I., & Zaharie, M. M. (2013). *New empirical insights into advertising creativity—traditional/non-traditional media context.* International Conference "Marketing—from Information to Decision" (6th ed., pp. 92–101). Risoprint Publishing House, Romania.

Krautsack, D. (2008). Ambient media-how the world is changing. *Admap Mag, 472*, 24–26. Retrieved from https://goo.gl/ByKeff.

Lee, Y. M. & Dacko, S. (2011). Ambient marketing: Towards a modern definition. In A. Patterson & S. Oakes (Eds.), *Proceedings of the Academy of Marketing Conference 2011 Marketing Fields Forever*, Liverpool: Academy of Marketing.

Lee, S., & Rao, V. S. (2010). Color and store choice in electronic commerce: The explanatory role of trust. *Journal of Electronic Commerce Research, 11*(2), 110–126.

Lehnert, K., Till, B. D., & Carlson, B. D. (2013). Recall, wearout and wearin effects. *International Journal of Advertising. The Review of Marketing Communications, 32*(2), 211–231.

Leung, L. (2004). Net-Generation attributes and seductive properties of the Internet as predictors of online activities and Internet addiction. *Cyberpsychology & Behavior, 7*(3), 333–348.

Levinson, J. C. (2003, November 17). Guerrilla marketing in a tough economy: To succeed during an economic rough patch, you have to think and act like a successful guerrilla marketer. *The Entrepreneur.* Retrieved from https://goo.gl/B2KKSG.

Lutz, R. J., MacKenzie S. B., & Belch, G. E. (1983). Attitude toward the ad as a mediator of advertising effectiveness. Determinants and consequences. *Journal of Marketing Research, 10*(1), 532–539.

Luxton, S., & Drummond, L. (2000, November 28–December 1). What is this thing called ambient advertising? *Proceedings of the Australia and New Zealand Marketing Academy Conference,* Gold Coast, Australia. (pp. 734–738). Retrieved from https://goo.gl/DoYYxD.

Mago, Z. (2015). Implicit in-game advertising: the tool of self-promotion and cross-promotion in digital games. *Analýza a výskum v marketingovej komunikácii, 3*(1), 30–38.

Maniu, A.-I., & Zaharie, M.-M. (2014). Advertising creativity—the right balance between surprise, medium and message relevance. *Procedia Economics and Finance, 15,* 1165–1172. doi: 10.1016/S2212-5671(14)00573-5

Mannheim, K. (1952). The problem of generations. In P. Kecskemeti (Ed.), *Essays on the sociology of knowledge* (pp. 273–323). London: Routledge & Kegan Paul.

McCracken, G. (1989). Who is the celebrity endorser? Cultural foundation of the endorsement process. *Journal of Consumer Research, 16*(1), 310–321. doi:10.1086/209217

McCrindle, M. (2006). *Understanding generation Y.* North Parramatta, Australia: The Australian Leadership Foundation. Retrieved from https://goo.gl/HII8Sg

McLuhan, M. (1991). *Jak rozumět médiím.* [The extension of man]. Praha, Slovakia Republic: Odeon.

Mehta, A. (2000). Advertising attitudes and advertising effectiveness. *Journal of Advertising Research, 40*(3), 67–72. doi:10.2501/JAR-40-3-67-72

Meyer, W. U., & Niepel, M. (1994). Surprise. In V. S. Ramachandran (Ed.), *Encyclopedia of human behavior* (Vol. 4, pp. 353–358). Orlando, FL: Academic Press.

Mikuláš, P., & Světlík, J. (2016). Execution of advertising and celebrity endorsement. *Communication Today, 7*(1), 93–103.

Mingroni, M. A. (2004). The secular rise in IQ: Giving heterosis a closer look. *Intelligence, 32*(1), 65–83.

Mistrík, M. (2004). Elektronické médiá [Electronic media]. In S. Magál & A. Pelcner (Ed.), *Kolokvium 1- 2 Katedry masmediálnej komunikácie [Colloquium 1-2 department of mass media communication]* (pp. 15–24). Trnava: Univerzita sv. Cyrila a Metoda v Trnave.

Nielsen. Trust in Advertising, a global Nielsen Consumer Report. (2013). *Nielsen global online consumer survey.* Retrieved from https://goo.gl/HKiE6M.

Oblinger, D. G., & Oblinger, J. L. (2005). *Educating the net generation.* An Educause e-book publication. Retrieved from https://goo.gl/g3TjVw.

Ogonowska, A. (2014). *Twórcze metafory medialne* [Creative metaphor of media]. Kraków: Universitas.

Ohanian, R. (1991). The impact of celebrity spokespersons 'perceived image on customers' intention to purchase. *Journal of Advertising Research, 31*(1), 46–54.

Pavel, C., & Cătoiu, I. (2009). Unconventional advertising for unconventional media. *Revista Economică, 2*, 142–145.

Pavlů, D. (2009). *Veletrhy a výstavy–kultura, komunikace, multimedialita, marketing* [Trade fairs and exhibitions–culture, communication, multimediality, marketing]. Praha, Slovakia: Professional Publishing.

Pavlů, D. (2013). Kde je ukryta komunikační specifika veletrhů a výstav a proč ji hledá jen tak málo marketérů? [Where are hidden the communication particularities of fairs and exhibitions and why are they so weakly seeked by marketers?] *Dot.comm. Journal for the Theory, Research and Practice of Media and Marketing Communication, 1*(1–2), 7–18.

Petty, R. E., & Cacioppo, J. T. (1986). The elaboration likelihood model of persuasion. *Advances in Experimental Social Psychology, 19*, 123–205. doi:10.1016/s0065-2601(08)60214-2

Postman, N. (1999). *Ubavit se k smrti.* [Amusing ourselves to death]. Praha, Slovakia: Mladá fronta.

Prensky, M. (2001). Digital natives, Digital immigrants. *On the Horizon, 9*(5), 1–6. Retrieved from https://goo.gl/YxIYNH.

Rábeková, R. (2008). *Afinita k médiám podľa cieľových skupín* [The affinity to the media by target groups]. Nitra, Slovakia: Univerzita Konštantína Filozofa.

Reinartz, W., & Saffert, P. (2013). Creativity in advertising: When it works and hen it doesn´t. *Harvard Business Review, 3*(6), 3-8. Retrieved from https://goo.gl/lSzww7

Rösler, H. D. (1990). Secular acceleration in psychological and somatic trends. A review. *Arztl Jugendkd, 81*(2), 76–85.

Rothenberg, A., & Hausman, C. R. (1976). *Introduction: The creativity question.* Durham, NC: Ed. Duke University Press.

Sankaran, S. (2013). A study on the food product commercials in television with reference to humour appeal. *Journal of Mass Communication and Journalism, 3*(4), 1-9. doi:10.4172/2165-7912.1000158. Retrieved from https://goo.gl/qJFYL8.

Saucet, M. (2015). *Street marketing TM: The future of guerrilla marketing and buzz.* Santa Barbara, CA: ABC-CLIO.

Scherer, H. (2004). Úvod do metody obsahové analýzy [Introduction to methods of content analysis]. In W. Schulz, I. Reifová (Eds.) *Analýza obsahu mediálních sdělení* [*Content analysis of media messages*] (pp. 29–50). Praha, Slovakia: Karolinium.

Shankar, A., & Horton, B. (1999). Ambient media: Advertising's new media opportunity? *International Journal of Advertising, 18*(3), 305–321.

Sheinin, D. A., Varki, S., & Ashley, Ch. (2011). The Differential effect of Ad novelty and message usefulness on brand judgment. *Journal of Advertising, 40*(3), 5–17. doi:10.2753/JOA0091-3367400301

Shimp, T. A. (2007). *Advertising, promotion, and other aspects of integrated marketing communications* (7th ed.). Mason, OH: Thomson South-Western.

Šimek, I., & Jakubéczyová, V. (2003). *Postoje verejenosti k reklame. Medzinárodná štúdia zo strednej Európy* [Public attitudes to advertising. International studies from Central Europe]. Taylor Nelson Sofres SK. Retrieved from http://www.tns-global.sk/docs/Postoj_k_reklame1.ppt.

Smith, R. E., Mackenzie, S. B., Yang, X., Buchholz, L. M., & Darley, W. K. (2007). Modeling the determinants and effects of creativity in advertising. *Marketing Science, 26*(6), 819–833.

Smith, J. W. (2006). Coming to concurrence: Improving marketing productivity by reengaging resistant consumers. In N. S. Jagdish & S. S. Rajendra (Eds.), *Does marketing need reform? fresh perspectives on the future* (pp. 15–25). New York, NY: M. E. Sharpe.

Soh, H., Reid, L. N., & King, K. W. (2007). Trust in different advertising media. *Journalism and Mass Communication, 84*(3), 455–476. doi:10.1177/107769900708400304

Solík, M. (2014). Semiotic approach to analysis of advertising. *European Journal of Science and Theology, 10*(1), 207–217. Retrieved from https://goo.gl/rmeyB5.

Štrbová, E. (2012). *Organizácia a motivácia v event marketing* [Organization and motivation in event marketing]. Nitra, Slovakia: Univerzita Konštantína Filozofa.

Survio. (2017). Online survey [Online software]. Retrieved from http://www.survio.com/sk/.

Světlík, J. (2012). *O podstatě reklamy* [About essence of advertising]. Bratislava, Slovakia: Eurokódex.

Szabo, P., & Szabová, E. (2014). Experience with project development in teaching statistics for mass media studies. *Studia Ekonomiczne. Marketing Communication–Selected Issues, 14*(205), 133–144.

Szobiová, E. (2004). *Tvorivosť–od záhady k poznaniu. Chápanie, zisťovanie a rozvíjanie tvorivosti* [Creativity-from mystery to cognition. Understanding, identifying and developing creativity]. Bratislava, Slovakia: Stimul.

Szyszka, M. (2013). *Kształtowanie wizerunku instytucji pomocy społecznej w mediach* [Forming the image of social assistance institutions in the media]. Warszawa, Poland: Centrum Rozwoju Zasobów Ludzkich.

Tapscott, D. (2009). *Grown up digital: How the net generation is changing your world.* New York, NY: McGraw-Hill.

Tellis, G. J. (2000). *Reklama a podpora prodeje* [Advertising and sales promotion]. Praha, Slovakia: Grada.

Tomczyk, Ł. (2010). Seniorzy w świecie nowych mediów [Seniors in the world of new media]. *E-mentor, 4*(36), 52–61. Retrieved from https://goo.gl/W61xe7.

Tripp, C., Jensen, T. D., & Carlson, L. (1994). The effect of multiple product endorsements by celebrities on consumer attitudes and intentions. *Journal of Consumer Research, 20*(4), 535–547. doi:10.1086/209368

Um, N. H. (2008). Exploring the effects of single vs multiple products and multiple celebrity endorsements. *Journal of Management and Social Sciences, 4*(2), 104–114. Retrieved from https://goo.gl/lMx62d.

Vrabec, N., & Petranová, D. (2013). *Nové vzory mládeže v kontexte mediálnej komunikácie* [Youth new models of in the context of media communication]. Trnava, Slovakia: Univerzita sv. Cyrila a Metoda v Trnave.

Vybíral, Z. (2005). *Psychologie komunikace. Psychologické aspekty masmediální komunikace* [Psychology of communication. Psychological aspects of mass media communication]. Praha, Slovakia: Portál.

Walotek-Ściańska, K. (2015). *Teatry publiczne w województwie śląskim a social media* [Public theaters of Silesia province and social media]. Katowice, Poland: Wydawnictwo Naukowe Śląsk.

Wąsiński, A. (2005). Autokreacja człowieka w dobie kultury obrazkowej jako nowy aspekt badawczy pedagogiki mediów [Autocreation man in the era of culture of image as a new aspect of research media pedagogy]. *Neodidagmata, 27-28,* 53–66. Retrieved from https://goo.gl/gYnQ42.

Wells, W. D., Burnett, J., & Moriarty, S. (1965). *Advertising principes and practice.* New York, NY: Prentice Hall.

Wilczek, P., & Fertak, B. (2004). Ambient media, media tradycyjne–konkurencja czy współpraca? [Ambient media, traditional media—competition or cooperation?]. *Brief, 52, 7-12.*

Wilson, R. T., Baack, D. W., & Till, B. D. (2011). Outdoor advertising recognition affects: Attention and the distracted driver. *Lubbock: American Academy of Advertising. Conference. Proceedings,* Lubbock, TX, 162–163. Retrieved from https://goo.gl/jnQFJX.

Wojciechowski, Ł. P. (2014). Ambient marketing as new language of marketing communication and an original use of the Urban space. *Annales Universitatis Paedagogicae: Studia Poetica II, 2*(169), 134–146.

Wojciechowski, Ł. P. (2016). *Ambient marketing + case studies in V4.* Kraków, Poland: Towarzystwo Słowaków w Polsce.

Chapter 10

Social Media and Trust

Jerzy Gołuchowski, Barbara Filipczyk, and Joanna Paliszkiewicz

Contents

Introduction

In social media, trust and its analysis are crucial, especially in this age of Big Data (Calefato, Lanubile, & Novielli, 2015; Niedermeier, Wang, & Zhang, 2016; Wendy Chan, 2015). Both insecurity and the variability of choices make trust a significant condition for cooperation within each society, including a virtual one. However, the increasing specialization of roles, functions, and professions make such cooperation essential, and also place this cooperation as another source of insecurity (Sobocińska, 2009). Therefore, trust becomes an instrument that facilitates both the evaluation and selection from a wide variety of possibilities (cf. Sztompka, 2007), as well as reduces insecurity.

The issue of trust in social media is complex, not only due to the factors mentioned above, but also because of attitudes toward truth in media coverage in the modern world (which includes the social media world). Aside from the trend considering truth as a basic evaluating criterion, there is a strong trend that recognizes post-truth as a dominant feature of the present. According to the basic meaning presented in the Oxford Dictionary, post-truth is defined as "relating to or denoting circumstances in which objective facts are less influential in shaping public opinion than appeals to emotion and personal belief" (Oxford Dictionaries, 2016). The notion accents that the beliefs of people today, including social media users, are fueled by emotive opinions and arguments rather than facts. For instance, the concept was used to capture the gut instinct, anti-establishment politics that swept Donald Trump and Brexit supporters to victory. In both campaigns, social media played a significant role.

Trust is understood and defined in various ways (Paliszkiewicz, 2013; Paliszkiewicz, Gołuchowski, & Koohang, 2015; Taddeo, 2009; Tejpal, Garg, & Sachdeva, 2013). The difficulties associated with the creation of universal trust definitions result from the complexity of the trust phenomena and, in consequence, from the research interdisciplinary (Kim, Ferrin, & RaghavRao, 2009). The issue of trust has been the subject of research conducted within the fields of sociology, political sciences, philosophy, psychology, economy, management sciences, and media sciences (Curtis, Herbst, & Gumkovska, 2010; Gauchat, 2012; Muethel & Bond, 2013; Murphy, 2006; Park, 2015; Rousseau, Sitkin, Burt, & Camerer, 1998; Ward, Mamerow, & Meyer, 2014). This is confirmed by the analyses of numerous works on this topic.

In the early scientific studies on trust, its definitions concentrated mainly on its psychological aspects. It was only later that the notion was considered in a broader context, aiming to create a universal definition. However, the multitude and dissimilarity of both trust definitions and the measurement models hampered the development of a common approach to the various authors' work results, thus making these results difficult to compare. Currently, the trend to particularize this notion is visible again. It is analyzed, for instance, in the context of the virtual environment, in the context of trust toward new fields, for example, technologies or institutions as well as in terms of trust in social media (Paliszkiewicz & Koohang, 2016).

Trust, trust models, and trust management systems have been researched more in recent years, but in the field of social media analysis, the lack of its unambiguous definition is still visible. It is necessary to capture the knowledge of trust as a foundation of an automatic analysis.

Satisfactory answers have not been found, not only to epistemological questions focusing on the research on trust in social media, but also to methodological ones. The research methods that should be utilized to obtain a better empirical experience of trust (Calnan & Rowe, 2006) or the methods of trust management are not established, yet. The generally accepted methodology of trust measurement, which could constitute a foundation of the analytical framework for building and diagnosis the trust in social media, seems necessary.

The aim of this chapter is to capture the characteristics of trust in social media through the development of trust ontology in the social media context. This seems to be a step that needs to be taken in terms of the abovementioned social media trust measurement. The development of ontology creates a foundation for the analytical framework aiming to build and diagnose trust in social media presented in this chapter.

Trust in Social Media and Trust-Based Knowledge Analysis

Social Media

Social media revolutionized communication in today's world by creating a new communication area—the Internet (also known as cyberspace or virtual reality) (Hadzialic, 2016; Hynan, Murray, & Goldbart, 2014; Tench & Jones, 2015). This permits an almost simultaneous communication between nearly unlimited institutions online which are territorially dispersed. Therefore, social media constitutes a platform of mutual contact of global network users situated anywhere in the world. Such platforms enable the exchange of thoughts, sharing experiences, concurrent sharing of users' information to a group of people (e.g., their location or emotional state), and allow pictures and sharing to the user's list of friends. They create an area where one can create their own image, seek social support, organize social events, or undertake political activity. For instance, due to quick communications from anonymous users from different countries (Anonymous) are able to organize protest movements.

The world of social media has also been penetrated by institutions and companies that use it as a means of sharing information and contacting their clients (Ngai, Moon, Lam, Chin, & Tao, 2015). Company profiles and known brands are often added to the user's friend lists due to their interest in the companies' products.

Many individuals treat the information obtained from social media—portals or blogs—as more important than the ones available in the traditional means of communication. In this way, social media has resulted in sharing opinions, knowledge, and public or private discussions. Several types of discourses have emerged and created many implications on the private, institutional, social, and political activities. They have an impact on communication, as well as on the work in many institutions and companies. Medialization through social media has created new, innovative forms of online interaction and contribution and has opened up new perspectives for knowledge acquisition and sharing as well as participation in the decision-making processes (Bharati, Zhang, & Chaudhury, 2015).

The notion of social media encompasses forms of communications such as blogs, microblogs, wikis, social networking sites, and content sharing sites (for further info see: Paliszkiewicz & Koohang, 2016). The benefit to businesses that have access to a communication platform for a targeted audience has been broadly described in many publications. It is often suggested that social media offers significant benefits for corporate communication. The benefit of social networking using a variety of social media enable discourse, knowledge distribution, and creation of ideas without being in close proximity physically. As a means of communication, its use can reduce the impact of the disability. For example, a visually impaired individual may use social media, for example, Facebook or LinkedIn, to communicate with friends and people working within the same area of scientific research.

Trust as a Relation

The relationship between partners communicating by social media is based on trust (de Laat, 2005). It involves the engagement of at least two parties participating in the relationship: the trusting party (the trustor) and the trusted party (the trustee). The characteristics of social media include the fact that each of the participants (parties) of the communication process can, in a certain moment, act as a trustor, and as a trustee in the next one (or inversely). In social media, the trustor signals their credibility first through the creation of a post in the social media sphere—thus becoming the trustee/potential trustee (see Figure 10.1). Credibility and other trust foundation elements are signaled by various means in a message being subject to a credibility assessment.

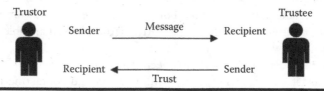

Figure 10.1   Relationship between the trustor and trustee.

The foundations of the relationship, where the sender's credibility is significant and which is built on trust (i.e., foundations that may be difficult to capture at times), are shaped by the final level of the message recipient's trust toward this message and its sender. Such trust comes to life between the partners of the peculiar relationship, created between the message sender and message recipient. It can be metaphorically stated that the sender "sells" the content and the readers, through its acceptance and acquisition (by making it "theirs") become the buyers. The signals contained in the text and its context are decoded by the individual who endows the sender with trust. In the last stage, based on the signals and other circumstances (psychological tendency to trust), the person credits the trustee with a certain amount of trust (Doligalski, 2009).

In summary, the approach toward trust presented in the literature shows that it is treated as, among others, a collection of determined expectations toward the partner, a consent to remain in a co-dependence or dependence state, a trustee's belief that their expectations regarding the delivery of certain values from the trustor are met (and that they will not be at the risk of additional costs) (Doligalski, 2009), or a peculiar relationship between the partners in the communication process.

What seems especially interesting is the approach toward trust as a relationship because it allows understanding the characteristic features that accompany the creation of the trust. Trust as a relationship between communication partners has many features, which are characterized in Table 10.1.

Table 10.1 Features of the Trust Relationship

Trust Relationship Feature	Description
Trust is not symmetric	The trusting person/agent does not necessarily have to be endowed with trust by the relationship partner. If "A trusts B," it does not follow that "B trusts A."
Trust is not distributive	If "A trusts (B and C)," it does not follow that "(A trusts B) and (A trusts C)."
Trust is not associative	The trust-operator does not map from entities to entities. "(A trusts B) trusts C" is not a valid trust expression. However, "A trusts (B trusts C)" is a possibility.
Trust is not inherently transitive	If "A trusts B" and "B trusts C," it does not automatically follow that "A trusts C."

Source: Viljanen, L., Towards an ontology of trust, *Proceedings of the 2nd International Conference on Trust, Privacy and Security in Digital Business (TrustBus'05)*, Copenhagen, Denmark, 2005.

Trust is a complex phenomenon and interesting studies are associated with its foundations. They are referred to as trust elements (Daniel & McAllister, 1995), trust dimensions (Erdem & Özen-Aytemur, 2014; Paliszkiewicz, 2013; Tan & Lim, 2009), trust sources (Zejda, 2010), or trust components (Kramer & Tyler, 1996). In the associated literature, the term "trust dimensions" is used interchangeably with other notions in the analyses which vary in terms of the level of detail.

Trust Components

Trust is a result of both, cognitive/epistemological processes, related to knowledge acquisition, and psychological ones, where a significant role is allocated to foundations and emotions (affective/emotional aspects). Bestowal of trust is, therefore, a complex evaluation of various characteristics constituting trust (including the evaluation of the relationship partner), conducted by one of the parties. The evaluation is not unchangeable as there are many processes associated with trust such as trust development, trust preservation, loss of trust, trust re-build/re-development, and so on.

Throughout the years of trust studies, the researchers have not reached an agreement regarding which of the trust features (dimensions, components, elements), numerously presented in the literature, are the most significant. The diversity of the research results available in the literature in this field is great. It includes theoretical models and empirical research where the researchers propose as many as 10 dimensions (McCole, 2002), but there are also research works assuming only two components: affect-based trust/goodwill-based trust and cognition-based trust/knowledge-based trust (Korsgaard, Schweiger, & Sapienza, 1995; Lewis & Weigert, 1985; McAllister, 1995; Whitener, Brodt, Korsgaard, & Werner, 1998; Wicks, Berman, & Jones, 1999; Young & Daniel, 2003). The distinction of these components is especially popular among the researchers concerned with trust measurement. Most scientists assume that trust includes both of the elements (McAllister, 1995), however, some of them claim that it is essentially an affect-based reaction (Atkinson & Butcher, 2003; Korsgaard et al., 1995). Undoubtedly, visibility is the influence of the beliefs related to the relationships between knowledge/cognition and the affective/emotive side (or between the cognitive and psychological processes) on the shape of these notions. For instance, the influence of emotions on the cognitive processes is visible.

Affect-based trust is associated with the emotive side of the relationship between partners. Flores and Solomon (1998) claim that trust based on the affection/emotions has an entirely different character than trust based on rational premises. It is emphasized that this form of trust requires a common interest, open communication, integrity, the experience of positive interactions, good intentions, attention to the relationship upholding (Lencioni, 2012), and that it requires more time to emerge (Gulati & Sytch, 2008; Pučėtaitė, Lamsa, & Novelskaite, 2010).

Cognition-based trust is related to the trusting party's knowledge of the party endowed with trust. Knowledge takes the shape of a conviction resulting from, above all, a certain experience acquired in the course of the relationship with the partner. Many researchers, when distinguishing the so-called cognitive trust (see e.g., Tyler & Kramer, 1996), find support in the forecasts and calculations associated with the likelihood of the occurrence of the second party's given behavior. This approach toward trust implies that the premises of trust toward the other party include the knowledge on the similarity of moral and ethical beliefs of the parties, the acquaintance of their hitherto actions, as well as the knowledge that both parties acted in a credible, competent and responsible way. Due to this knowledge, the parties may also expect a similar behavior in the future. This makes the cognitive trust to be supported by rational premises. Cognition-based trust relies on the perception of credibility, which is mostly conditioned by the evaluation of the second party's competence, kindness, and integrity (Mayer, Davis, & Schoorman, 1995).

The distinction of all factors influencing the process of trust shaping in the social media is highly difficult. Trust is developed by both honest individuals and manipulators. It is possible to learn how to build trust, therefore the analyses of reports, studies of documents and statements available in the social media, permitting capturing good practices, are justifiable.

Developing Trust in Social Media

Trust in interpersonal and inter-organizational relationships is a resource that usually enhances with the development of contacts and diminishes with their lack. However, this relationship is not developed automatically. The development of trust depends on many factors associated with both sides of the relationship. From the perspective of the individuals who attribute a large significance to trust, including the managers in organizations utilizing social media to communicate both within the company and with its environment, the ability to "direct" the trust development process is of great importance. It involves the creation of appropriate circumstances for trust stimulation and the systematic monitoring of its level as well as possible updates of the knowledge of the relationship partners and the planned actions. Therefore, members of the organizations who are managing social media communication should be aware that mature trust requires time and a transition through several intermediate stages. What may be of use in this aspect is a tool for trust monitoring and shaping.

What is worth noticing and including both in the trust development process and in the trust analyses is the fact that trust building must be gradual and should be based on the delivery of new evidence to strengthen trust. Initial trust is developed on the overall trust manifesting in the way which each of the partners treats other individuals. However, full trust with which a partner can be endowed in the communication relationship is a process requiring time and the acquisition of new knowledge on the media communication partners. The research shows that a conviction, which is an effect of the second party's credibility evaluation resulting from rational or emotive premises, is needed prior to and aside from the overall trust. A conviction build in this way induces the decision to engage in the relationship and undertake further effort aiming to strengthen the initial trust or even enhance it. The decision and its subsequent actions may also result from rational or emotive premises.

When it comes to organizations, the trust management term is also used (Paliszkiewicz, 2013). Trust management is defined as an activity based on the creation of systems and methods which enable both parties to evaluate the trust level and make decisions whether to engage in the potential transaction and to estimate the level of the risk associated with it. Trust management is a more voluminous notion than trust development, as it also involves other aspects. It includes the activities associated with trust, whereas trust building is focused on only one of the trust management processes.

In the trust management process, the organizations not only build (develop) trust using social media but also struggle with other, day-to-day issues such as trust erosion, trust re-build, the coexistence of trust, and distrust or trust transfer. All of these notions should remain within the range of the organization's management of trust, developed using the social media. Therefore, this allows the creation and analysis of the conditions (processes, principles) in which trust develops and occurs.

Levels of Trust Maturity

The researchers identified **five maturity levels** associated with trust and dependent on the quality and phase of relationship development. The characteristics of these levels are presented in Table 10.2.

For instance, the development of **knowledge-based trust**, according to Lewicki and Tomlinson (2003), constituting a more permanent category of trust, requires actions, including knowledge acquisition and experience gaining, due to the fact that the previous experience and the knowledge gathered on the partner (especially the predictability of their behavior) are in its foundations. The creation of this kind of trust requires also a necessary factor of a positive contact history confirming the partner's credibility. This type of trust is developed in the course of effective communication between the parties as it deepens the mutual acquaintance and understanding between them. The relationships in social media may be based on this kind of trust, as they are related to partners knowing each other for a certain period of time. It is built on the experience of a successful cooperation and the verification of mutual expectations.

It is worth noting that Lewicki and Tomlinson (2003) deemed **identification-based trust** as the highest level of trust, with its foundation defined as an emotional bond that generates mutual loyalty. This bond emerges due to the parties'

Table 10.2 The Characteristic of the Trust Maturity Levels

Level	Level Name	Characteristics	Type of Trust
1	Distrust	This level does not meet the definitional conditions of trust due to the lack of positive expectations and good will of the involved parties. Sanction threats are a characteristic property of this trust level.	**Deterrence-based trust**
2	Low trust	It can be described as a strategy based on a very precise analysis of the costs and benefits associated with the relationship. It does not fully involve trust features.	**Calculus-based trust**
3	Confident trust	It meets the definitional criteria of trust and its characteristics comply with the majority of the definitions proposed by the researchers. This trust appears when the suspicion transforms into positive expectations toward the relationship partner, their motives, abilities, or integrity.	**Knowledge-based trust**
4	Strong/ high trust	This level of trust appears when the expectations set for the partner are confirmed by the experiences. The level has an affective character. It is a result of the relationship quality rather than the observations of the other party's specific behavior.	**Relational-based trust**
5	Complete trust	Strong emotional bonds as well as the unity of interests and objectives support the creation and lead to the acceptance of a common identity. Each of the trust relationship partners can represent the common interests with full trust.	**Identification-based trust**

Source: Dietz, G., & Den Hartog, D. N., *Personnel Review*, 35(5), 557–588, 2006.

very good knowledge of each other as well as the ability to represent each other's interests. This trust relationship is symmetric. Control and security are necessary for consequence of this level of trust maturity. Such a high level of trust may occur in organizations and in the social media only in specific cases—in very well integrated, small teams.

Analytical Model of Trust in Social Media

When managing trust, the managers responsible for the communication and knowledge management perform at least two key tasks:

1. A diagnosis of trust in the community that the organization is interested in/the determination of the level of the maturity of the trust toward the organization
2. Evaluation of the "quality" of the trust-based knowledge created in the group

During the analysis of the discourse in social media, there is a need to identify the type of trust granted by the partners fulfilling the roles of the trustor or the trustee as well as to evaluate the various types of the trust-based knowledge. Moreover, there is a need to evaluate trust itself due to the fact that in a group utilizing social media as a means of communication and for trust-based knowledge creation, trust may be on a different maturity level. Therefore, there is a need for a tool to diagnose the existing trust as well as a tool used to evaluate the knowledge delivered by the participants of the discourse in the social media. A robust understanding and trust analysis, especially automatic, is permitted by a good ontology of trust in social media.

Trust Ontology in Social Media

The literature features numerous definitions of this notion. The broadly accepted definition of ontology is a subject of an ongoing discussion. The definition formulated by Gruber in 1993 is considered the most commonly known. According to this definition, "An ontology is a formal, explicit specification of a shared conceptualization" (Gruber, 1993), the conceptualization is an abstract, simplified representation of the world which is desired to be presented for a specific purpose. "Explicit specification" emphasizes the role that notions and relationships occurring in the abstract reality model as well as their limitations, should be assigned with unambiguous names and definitions. "Formal specification" of ontology indicates that the definition of the terms and notions should be created in a formal language with well-known logic-based properties. Shared conceptualization means that ontology may be used by many agents; humans and applications.

According to W3C, it is assumed that "An ontology defines the terms used to describe and represent an area of knowledge. Ontologies are used by people, databases, and applications that need to share domain information (...). Ontologies include computer-usable definitions of basic concepts in the domain and the relationships among them (note that here and throughout this document, definition is not used in the technical sense understood by logicians). They encode knowledge in a domain. In this way, they make that knowledge reusable." (W3C Recommendation, 2004).

The methodology of ontology creation has been presented in numerous research works (i.e., Casellas, 2011; Corcho, Fernandez-Lopez, & Gomez-Perez, 2006; Gomez-Perez, Fernandez-Lopez, & Corcho, 2004).

Trust Ontology Assumptions

When creating an ontology, it is necessary to determine its objectives and accept the assumptions it needs to be fulfilled. It is worth mentioning that a trust-based relationship between the partners in social media includes the involvement of at least two parties participating in the relationship: the trusting party (trustor) and the trusted party (trustee). The subject literature distinguishes two approaches to the issue of the determination of the roles associated with the creation of a trust-based relationship.

The first approach, derived from psychology, assumes that trust is deeply rooted in the personality of the trustor and its strength is dependent on the individual psychosocial development (Rotter, 1967). This trust is sometimes referred to as affective trust. This approach assumes that trust depends mostly on the trustor. It emphasizes that trust, while being on the trustor's side, results from the person's readiness to trust other people.

The second approach is based on the assumption that trust is dependent on the trusted subject. The subject does not imply a person—it may be a brand, a technology, or an institutional system. Trust is a positive expectation related to the motives of the second party associated with the evaluation of one's own risk level (Lewicki & Bunker, 1996). According to this approach, trust mostly depends on the relationship partner's ability to develop credibility (Laeequddin, Sahay, Sahay, & Waheed, 2010). Usually, the trusted party's trust is considered.

It seems that a more complex epistemological approach toward trust is needed for its development and analysis in the social media: a trust ontology taking into account not only the attitudes of the relationship partners but also the trust subject and context. What needs to be examined is the trustor—trust subject—trustee relationship in the given time as well as the trust decision-making process set in the particular case. This is confirmed by the results of numerous studies proving that trust consists of many dimensions. It contains not only the mental and emotional component/aspect but also the cognitive one. In such works, trust is understood as a certainty, a conviction or a conviction collection, a hope or the confidence in the future behavior of the other party.

Social Media Trust Ontology

The top level of the proposed social media trust ontology is presented in Figure 10.2.

a. Major classes and subclasses of social media trust ontology
b. Class Actor and subclasses

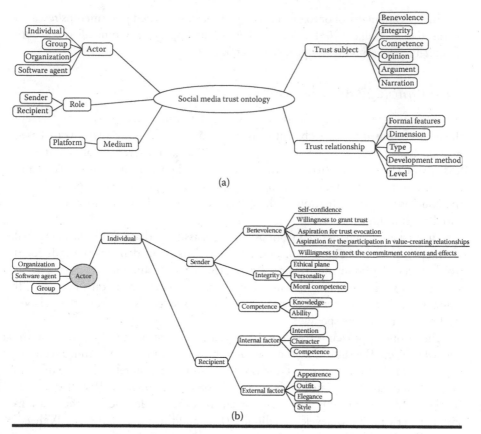

Figure 10.2 Top level social media trust ontology. (a) Major classes and sub-classes of social media trust ontology. (b) Class Actor and subclasses.

Medium

The conducted research shows that trust in social media depends on the social media platform hosting the discourse. Therefore, the ontology should also take into account the characteristics of this medium. The foundation of this characteristic constitutes in the description of social media (i.e., presented by Paliszkiewicz & Koohang, 2016).

Trust Sender Role

Self-knowledge, that is, the knowledge of the trusting subject of itself should be captured in trust ontology. It is important to take into consideration the following dimensions describing the trusting person: whom do they trust, how much do they trust, and why do they trust, including the emotional component (benevolence) and the cognitive component (the past behavior and its consequences, learning).

The willingness to trust (to grant trust) and its opposite, that is, distrust, are considered personality traits. The psychological perspective—the willingness to trust others shows that trust is a degree to which "one is willing to ascribe good intentions to and have confidence in the words and actions of other people. This willingness will in turn affect the way in which one behaves toward others" (Cook & Wall, 1980).

The willingness to trust others is also associated with the positive convictions of oneself. Persons declaring self-confidence are characterized as not only highly agential, that is, having the ability to deal with important life problems, but also as open-minded and trusting due to which they obtain the assistance and support from the people in their closest circle. They are also optimistic toward their future, seldom experience depression, sadness, and loneliness, and enjoy life more. Furthermore, they are more likely to learn new things. However, the role of the trusting party is not reserved for people only. In the media in general, the role is fulfilled by the message sender, but also by the social group (journalists), the organization (e.g., the newspaper editorial office), or even the Internet platform designed and ran by people. Recent studies show that a significant portion of social media traffic is generated by social bots.

Actors

The role is not always fulfilled by a human being—it can also be fulfilled by an organization, a company, or an institution (as an entity designed and ran by people). Trust may contribute to control action reduction between sides of media discourse in the same way as it can reduce uncertainty accompanying mutual business relations. It is based on a subjective, gradable likelihood of the occurrence of behavior and actions conducted by the other party in the particular situation. The level of trust is, therefore, a criterion deciding on the choice from a multitude of possibilities not only in the economic but also in media activity. In a situation where an action verifying the message needs to be taken, despite the risk, trust becomes the basic strategy of handling the uncertainty and the inability to control the future. Trust constitutes the belief in the given actions or properties of the trusted subject, it is a "bet taken in terms of uncertain, future actions of other people" (Sztompka, 2007, p. 69). The acceptance of the assumption indicates the expectations and convictions on a certain subject and the implementation of the convictions through actions. According to the quoted author, instead of passively awaiting what fate will bring, a person bases their convictions on the level of trust he or she has toward the other party. As a result, he or she undertakes actions based on trust, whose consequences are uncertain and impossible to predict.

Situation and Trust Subject

Trust depends on, not only the features of the trust relationship parties, but also on its subject, the content of the message and the situation in which the message is conveyed. Due to this fact, in many author's studies, trust is treated as a both situational and interpersonal variable. The subject of trust may include the sender's intentions (benevolence), the message sender's competencies, the compliance of actions with the declarations (integrity) and the products of the communication process: opinions, arguments, and narration.

Trust Relationship

The trust relationship is also treated as a separate entity when it is equipped with certain features. Their occurrence and intensity, and not only the relationship parties' features, are worth distinguishing and analyzing in the social media communication processes. In the associated literature, they are often attributed to the partners/trust parties.

The Prototypical Model

The outlined analysis of trust development and the aforementioned ontology constitute the foundations for the creation of the tool supporting the diagnosis of trust in relationships between partners who communicate via social media. The creation of the prototypical, analytical tool was based on the model developed by Lewicki and Tomlinson (2003) who indicated the actions one can undertake to build trust. The quality of the realization of these actions should be the subject of the diagnosis. The prototypical model of the tool is presented in Tables 10.3 and 10.4 and focuses on two levels of trust maturity: knowledge-based trust (the third maturity level) and identification-based trust (the highest level of trust maturity). These levels were characterized earlier in this chapter.

The first of the discussed trust levels—knowledge-based trust—can be diagnosed through the analysis of the communication experiences gathered in the course of the social media platform use. This provides foundations for gathering of knowledge on the media discourse partners, for example, determining the predictability of their behavior. This kind of analysis also requires the access to the positive/negative contact history occurring in the given time period, which would confirm the partner's credibility. This type of trust can be diagnosed when there is a permanent, effective communication between the parties.

The second of the presented trust levels—identification-based trust—is possible to obtain through the utilization of the analysis of the discourse participants' sentiment, as this level of trust is based on the emotional bond between the discourse partners. In their posts, the social media users refer not only to the discourse subject

Table 10.3 Knowledge-Based Trust

Analytical Construct	Characteristics	Measure of Quality
Fulfillment of the duties and tasks	Competent fulfillment of duties influences the credibility level.	Number of entries
Proficiency and professionalism in the fulfillment of duties	In certain cases, it may also indicate the update of the abilities and knowledge resulting from the technological progress.	Others' evaluations, number of positive marks
Attention to the consistency and predictability of actions	Increases the credibility level that is through redeeming a promise or the realization of obligations.	Number of corrections
Communication precision	Credibility increases due to the clarity associated with the intentions and motives of actions as both parties are willing to act in a transparent manner and undergo a possible control.	Evaluation of the attention paid to the language
Control reduction	Control is a sign of distrust and its limitation constitutes a proof of trust. According to the exchange theory, the granted trust returns, therefore the delegated control is of symbolic value.	Number of control actions
Caring for others	Trust increases if a sensitivity to the needs, desires, and interests of others is shown and if the desire to act only in one's interest is suppressed.	Number of given advice, acts of aid

but also voice their positive or negative opinions on the partners. These statements can, therefore, be used as the indicators of trust with which they endow their communication partners.

What seems to be helpful in the development of the trust analysis tool is the approach to trust as a peculiar process, proposed by Dietz and Den Hartog (2006). The authors distinguished three phases: belief, decision, and action. The adaptation of this model for the purpose of the analysis of the trust granted to the partner(s) of the trust relationship allows the assumption of three phases of the decision-making process.

The starting points (the first phase) of trust creation consists of the subjective judgment as well as the conviction related to the other party and the mutual

Table 10.4 Identification-Based Trust

Analytical Construct	Characteristics	Measure of Quality
Conditions of frequent interactions	Frequent interactions help both parties to acquaint themselves with each other better, strengthen their common identity, eliminate stereotypes and prejudices, and better demonstrate similarities rather than differences.	Number of interactions
The ability to create common products and achieve common goals	Establishing a common goal for the parties favors the sense of unity, community, and unusualness.	Feedback log
Caring for the common identity	Catering for the common identity creates the sense of unity that may be further enhanced through trust; involvement of both parties in the discussions and actions develop the sense of "us" instead of "I."	Content
Promotion of common values and the emotional attractiveness	Caring for others, listening, concentrating on the needs of other people and recognizing their contribution.	Content

relationship which assumes that the partner's actions will have positive consequences. In the case of an unfamiliar person, it is replaced by the overall trust—a trust toward a group of people that includes the particular person. On the other hand, in the case of a known individual, the trust can be calculated as the personal trust toward the given person.

The next phase, according to Dietz and Den Hartog (2006), focuses on the decision of whether to trust the relationship partner. According to many authors, it is located between the expectations associated with the credibility and the ones resulting from the previous actions (Huff & Kelley, 2003). The decision indicates that the trustor considers the trustee as credible on a certain level and intends to engage in the relationship, undertaking a given risk of the trustee's potentially harmful actions, determining their occurrence probability as rather low (Dietz & Den Hartog, 2006).

The last phase distinguished by Dietz and Den Hartog (2006)—the actions of the trustor directed at the partner—separates/joins the trust from/with the associated actions. The actions include control acts that the trustor may undertake (or not) in order to verify the evaluated message of the person who was assigned with a particular trust level. The verification delivers additional knowledge which may influence the earlier decision on the trust level. In the case of trust built by means of social media, the trust development process may include activities such as undertaking control actions that verify the analyzed opinion or argument through the search for other statements.

Conclusion

Social media is an important platform for the discourse which is participated in by, not only the individual social media users, but also the employees and organization associates. The objective of the discourse led by an organization is the acquisition and creation of credible knowledge used for innovations and decision-making. The development and analysis of trust in social media, in this particular context, is important in terms of creating and acquiring knowledge used by organizations for the purpose of decision-making.

This chapter presents the trust development and analysis concept utilizing the social media trust ontology and the outline of the tool that supported the manager in the diagnosis of trust within a group cooperating by means of the social media. The manager is interested in a group due to the interest in their trust toward the organization and the interest in the creation and acquisition of trust-based knowledge. It is assumed that in the case of trust built by means of social media, the entries on the users' activity, constituting the process of trust development, are at hand and their analysis allows the diagnosis of trust and in consequence, the evaluation of the quality of the knowledge (opinions, arguments, and narration) created with a group.

The ontology as well as the analytical tool presented in this chapter are prototypical and were verified using a relatively small group of journalism and information technology students of the University of Economics in Katowice. The verification has revealed the need to improve the developed tool and to develop a software tool (dashboard) for the analysis and presentation of the analysis results associated with the trust within a group utilizing social media to communicate, resolve problems, and create credible knowledge.

The deliberations conducted on trust demonstrated that many conceptually unresolved issues still occur. The created ontology allows the determination of the method of its understanding and generates a foundation for the development of trust analysis tools for the purpose of individuals dealing with communication and knowledge management within an organization. It seems that the main research issues associated with trust in social media, which still requires further study, includes the following:

- The character of social media trust in the aspect of its components, features (dimensions), and types
- The level of trust, especially its optimal level, for communication in social media
- Factors influencing trust developed by means of social media platforms, that is, creating and destroying factors
- The diagnosis of factors influencing the shaping of trust levels in social media
- The impact of the coexistence of trust and distrust between communication partners in social media
- The influence of new technologies (i.e., social bots) on the shaping of trust in social media
- Methods of measurement and automated analysis of trust in social media

Aiming to resolve these issues, further works of the authors will concentrate on the development of the methods of trust measurement in social media as well as on the construction of a tool for the automatic analysis of social media trust.

References

Atkinson, S., & Butcher, D. (2003). Trust in managerial relationships. *Journal of Managerial Psychology*, *18*(4), 282–304.

Bharati, P., Zhang, W., & Chaudhury, A. (2015). Better knowledge with social media? Exploring the roles of social capital and organizational knowledge management. *Journal of Knowledge Management*, *19*(3), 456–475.

Calefato, F., Lanubile, F., & Novielli, N. (2015). The role of social media in affective trust building in customer-supplier relationships. *Electronic Commerce Research*, *15*(4), 453–482.

Calnan, M., & Rowe, R. (2006). Trust relations in the 'new' NHS: Theoretical and methodological challenges, *Social Contexts and Responses to Risk Network* (SCARR). Working paper 14/2006, University of Kent at Canterbury Canterbury.

Casellas, N. (2011). Methodologies, tools and languages for ontology design. In N. Casellas (Ed.), *Legal ontology engineering. Methodologies, modelling trends, and the ontology of professional judicial knowledge* (pp. 57–107). Heidelberg: Springer.

Cook, J., & Wall, T. (1980). New work attitude measures of trust, organisational commitment and personal need non-fulfilment. *Journal of Occupational Psychology*, *53*, 39–52.

Corcho, O., Fernandez-Lopez, M., Gomez-Perez, A. (2006). Ontological engineering: Principles, methods, tools and languages. In C. Calero, F. Ruiz, & M. Piattini (Eds.), *Ontologies for software engineering and software technology* (pp. 1–48). Berlin: Springer.

Curtis, T., Herbst, J., & Gumkovska, M. (2010). The social economy of trust: Social entrepreneurship experiences in Poland. *Social Enterprise Journal*, *6*(3), 194–209.

Daniel, J., & McAllister, D. J. (1995). Affect- and cognition-based trust as foundations for interpersonal cooperation in organizations. *Academy of Management Journal, 38*(1), 24–59.

de Laat, P. B. (2005). Trusting virtual trust. *Ethics and Information Technology, 7*(3), 167–180.

Dietz, G., & Den Hartog, D. N. (2006). Measuring trust inside organizations. *Personnel Review, 35*(5), 557–588.

Doligalski, T. (2009). Budowa wartości klienta z wykorzystaniem Internetu [Building the customer value by using Internet]. In B. Dobiegała-Korona & T. Doligalski (Eds.), *Zarządzanie wartością klienta [Managing customer value]*, Warszawa, Poland: Poltext.

Erdem, F., & Özen-Aytemur, J. (2014). Context-specific dimensions of trust in manager, subordinate and co-worker in organizations. *Journal of Arts and Humanities, 3*(10), 28–40.

Flores, F., & Solomon, R. C. (1998). Creating trust. *Business Ethics Quarterly, 8*, 205–232.

Gauchat, G. (2012). Politicization of science in the public sphere: A study of public trust in the United States, 1974 to 2010. *American Sociological Review, 77*(2), 167–187.

Gomez-Perez, A., Fernandez-Lopez, M., & Corcho, O. (2004). Methodologies and methods for building ontologies. In A. Gomez-Perez, M. Fernandez-Lopez, & O. Corcho (Eds.), *Ontological engineering, with examples from the areas of knowledge management, E-Commerce and the semantic web* (pp. 107–197). London, New York, NY: Springer.

Gruber, T. S. (1993). A translation approach to portable ontology specifications. *Knowledge Acquisition, 5*(2), 199–220.

Gulati, R., & Sytch, M. (2008). Does familiarity breed trust? Revisiting the antecedents of trust. *Managerial and Decision Economics, 29*, 165–190.

Hadzialic, S. (2016). Transformation of the new communication media within the frame of interpersonal interaction. *International Journal on Global Business Management & Research, 5*(2), 116–134.

Huff, L., & Kelley, L. (2003). Levels of organizational trust in individualist versus collectivist societies: A seven-nation study. *Organization Science, 14*(1), 81–90.

Hynan, A., Murray, J., & Goldbart, J. (2014). 'Happy and excited': Perceptions of using digital technology and social media by young people who use augmentative and alternative communication. *Child Language Teaching and Therapy, 30*(2), 175–186.

Kim, D. J., Ferrin, D. L., & RaghavRao, H. (2009). Trust and satisfaction, two steeping stones for successful e-commerce relationships: A longitudinal exploration. *Information Systems Research, 20*(2), 237–257.

Korsgaard, M. A., Schweiger, D. M., & Sapienza, H. J. (1995). Building commitment, attachment, and trust in strategic decision-making teams: The role of procedural justice. *Academy of Management Journal, 38*(1), 60–84.

Kramer, R. M., & Tyler, T. R. (1996). *Trust in organizations: Frontiers of theory and research*. California, CA: Sage Publications.

Laeequddin, M., Sahay, B. S., Sahay, V., & Waheed, K. A. (2010). Measuring trust in supply chain partners' relationships. *Measuring Business Excellence, 14*(3), 53–69.

Lencioni, P. (2012). *The advantage: Why organizational health trumps everything else in business*. Hoboken, NJ: John Wiley & Sons.

Lewicki, R. J., & Bunker, B. B. (1996). Developing and maintaining trust in work relationships. In R. M. Kramer & T. R. Tyler (Eds.), *Trust in organizations: Frontiers of theory and research* (pp. 114–139). Thousand Oaks, CA: Sage Publications.

Lewicki, R. J., & Tomlinson, E. C. (2003). Trust and trust building beyond intractability. In G. Burgess, & H. Burgess (Eds.), *Conflict information consortium* . Boulder, CO: University of Colorado. Retrieved from http://www.beyondintractability.org/essay/trust-building.

Lewis, D. J., & Weigert, A. (1985). Trust as a social reality. *Social Forces, 63*, 969–985.

Mayer, R. C., Davis, J. H., & Schoorman, F. D. (1995). An integrative model of organizational trust. *Academy of Management Review, 20*(3), 709–734.

McAllister, D. J. (1995). Affect-and cognition-based trust as foundations for interpersonal cooperation in organizations. *Academy of Management Journal, 38*, 24–59.

McCole, P. (2002). The role of trust for electronic commerce in services. *International Journal of Contemporary Hospitality Management, 14*, 81–87.

Muethel, M., & Bond, M. H. (2013). National context and individual employees' trust of the out-group: The role of societal trust. *Journal of International Business Studies, 44*(4), 312–333.

Murphy, J. T. (2006). Building trust in economic space. *Progress in Human Geography, 30*(4), 427–450.

Ngai, E. W. T., Moon, K. K., Lam, S. S., Chin, E. S. K., & Tao, S. S. C. (2015). Social media models, technologies, and applications. *Industrial Management & Data Systems, 115*(5), 769–802.

Niedermeier, K. E., Wang, E., & Zhang, X. (2016). The use of social media among business-to-business sales professionals in china. *Journal of Research in Interactive Marketing, 10*(1), 33–49.

Oxford Dictionaries. (2016). *Oxford dictionaries word of the year 2016*. Retrieved from https://en.oxforddictionaries.com/word-of-the-year

Paliszkiewicz, J. (2013). *Zaufanie w zarządzaniu* [Trust in management]. Warszawa, Poland: Wydawnictwo Naukowe PWN.

Paliszkiewicz, J., & Koohang, A. (2016). *Social media and trust: A multinational study of university students*. California, CA: Informing Science Press.

Paliszkiewicz, J., Gołuchowski, J., & Koohang, A. (2015). Leadership, trust, and knowledge management in relation to organizational performance: Developing an instrument. *Online Journal of Applied Knowledge Management, 3*(2), 19–35.

Park, H. (2015). Is trust in government a short-term strategic value or a long-term democratic value? A case study of three nordic countries and three east Asian nations. *International Review of Public Administration, 20*(3), 273–286.

Pučėtaitė, R., Lamsa, A., & Novelskaite, A. (2010). Organisations which have the strongest potential for high-level organisational trust in a low-trust societal context. *Transformations in Business & Economics, 9*(2)/(20), Supplement B, 318-334.

Rotter, J. B. (1967). A new scale for the measurement of interpersonal trust. *Journal of Personality, 35*(4), 651–655.

Rousseau, D. M., Sitkin, S. B., Burt, R. S., & Camerer, C. (1998). Not so different after all: A cross-discipline view of trust. *Academy Manage Review, 23*(3), 393–404.

Sobocińska, M. (2009). Zaufanie w relacjach instytucji kultury z podmiotami otoczenia [Trust in relations between cultural and environmental institution]. In L. Garbarski & J. Tkaczyk (Eds.), *Kontrowersje wokół marketingu w Polsce. Niepewność i zaufanie a zachowanie nabywców* [Controversy surrounding the marketing in Poland. Uncertainty and confidence and the behavior of buyers]. Warszawa, Poland: WAiP.

Sztompka, P. (2007). *Zaufanie: fundament społeczeństwa* [Trust: The foundation of society]. Kraków, Poland: Wydawnictwo Znak.

Taddeo, M. (2009). Defining trust and E-trust: From old theories to new problems. *International Journal of Technology and Human Interaction, 5*(2), 23–35.

Tan, H. H., & Lim, A. K. H. (2009). Trust in coworkers and trust in organizations. *Journal of Psychology, 143*(1), 45–66.

Tejpal, G., Garg, R. K., & Sachdeva, A. (2013). Trust among supply chain partners: A review. *Measuring Business Excellence, 17*(1), 51–71.

Tench, R., & Jones, B. (2015). Social media: The wild west of CSR communications. *Social Responsibility Journal, 11*(2), 290–305.

Viljanen, L. (2005). Towards an ontology of trust, *Proceedings of the 2nd International Conference on Trust, Privacy and Security in Digital Business (TrustBus'05)*, Copenhagen, Denmark.

Ward, P. R., Mamerow, L., & Meyer, S. B. (2014). Interpersonal trust across six Asia-Pacific countries: Testing and extending the "high trust society" and "low trust society" theory. *PLOS ONE, 9*(4), 1–10.

Wendy Chan, W. L. (2015). Exploring the influence of social interaction, pressure and trust in a social media environment on political participation: The case of occupy central in 2014. *Online Journal of Communication and Media Technologies, 5*(4), 77–101.

W3C Recommendation. (2004). *OWL web ontology language. Use cases and requirements.* Retrieved from http://www.w3.org/TR/webont-req/#onto-def.

Whitener, E. M., Brodt, S. E., Korsgaard, M. A., & Werner, J. M. (1998). Managers as initiators of trust: An exchange relationship framework for understanding managerial trustworthy behavior. *Academy of Management Review, 23*(3), 513–530.

Wicks, A. C., Berman, S. L., & Jones, T. M. (1999). The structure of optimal trust: Moral and strategic implications. *Academy of Management Review, 24*, 99–116.

Young, L. C., & Daniel, K. (2003). Affectual trust in the workplace. *International Journal of Human Resource Management, 14*(1), 139–155.

Zejda, D. (2010). From subjective trust to objective trustworthiness in online social networks: Overview and challenges. *Journal of Systems Integration, 1*(1–2), 16–22.

ANALYTICS

Chapter 11

Advanced Analytics in Decision-Making

Alan Briggs

Contents

Part 1—Why You Should Do Analytics

Introduction

Analytics has been a frequent topic in major media outlets over the past several years. Many practitioners and academics alike argue that there is nothing new about the various technologies. In fact, many of them may have been around for some time. However, the new media attention has caught the attention of business executives, civic leaders, and stakeholders from nearly every corner of the globe. Everyone wants to be better at what they do, and analytics is offering an opportunity for many organizations to do just that. Building on a seemingly simple question, "What is analytics?" we ultimately seek to answer the question, "Why should you do analytics?"

What Is Analytics?

There has been a rousing debate within the analytics community over the past five or so years. As quantitative *analytical* techniques have garnered increased media attention and public interest in recent times, disciplines such as operations research, optimization, computer science, machine learning, artificial intelligence, mathematics, statistics, engineering (and on, and on, and on) have endeavored to pivot into *advanced analytics*. That is, everyone that has done anything *analytical* has attempted to articulate how their work fits into the field of *analytics*, and, reciprocally, how analytics fits into their field.

The term "analytics community" may be taking a bit of license, as there is not a single established definition of *analytics*, yet. And, perhaps there will never be. However, there are some well-established players in the field that have articulated their own definition, and exploring these could be useful to putting the pieces together. At least for the sake of the discussion at hand, a definition of analytics will be helpful in building up a tool set for leveraging advanced analytics in decision-making.

Here are four definitions worth introducing:

1. Analytics is an encompassing and multidimensional field that uses mathematics, statistics, predictive modeling and machine-learning techniques to find meaningful patterns and knowledge in recorded data—SAS (2017).*
2. Analytics is defined as the scientific process of transforming data into insight for making better decisions—INFORMS (2017).†

---

* https://www.sas.com/en_us/insights/analytics/what-is-analytics.html/, Retrieved March 1, 2017.
† https://www.informs.org/About-INFORMS/What-is-Analytics"/, Retrieved March 1, 2017.

3. Analytics is about asking questions from data and getting answers. It's what Aristotle and the ancient Greeks called "inquiry," and its two parts are the question being asked and the method of reasoning used to answer the question—Elder Research (2016).*
4. Advanced Analytics is a grouping of analytic techniques used to predict future outcomes—IBM (2014).†

These four definitions are from a diverse set of practitioners in advanced analytics, and each are undisputed leaders in the field. Consider these facts in establishing credibility for these sources:

■ The SAS Institute has been a leader in analytic software development for over 40 years, with installations in over 90 of the Fortune 100, and 90% of the Fortune 500 companies. They lead the marketplace with nearly one-third of the software revenue in advanced and predictive analytics, owning more than twice the market share of their closest competitor.

■ The Institute for Operations Research and Management Science (INFORMS) is the leading professional society for Analytics Professionals and Operations Researchers. They publish 14 scholarly journals that describe the latest O.R. and analytics methods and applications; organize conferences for academics and professionals; and provide analytics certification and continuing education to assist members and others in furthering their careers.

■ Elder Research is an experienced consultancy specializing in data science and text analytics solutions. They provide analytics consulting support throughout the journey—from training and mentoring, to creating stand-alone analytic solutions, to helping build the capacity to develop solutions in-house. They solve business problems as well as data problems.

■ IBM has been in the computing and technology business for over 100 years. They have assembled a broad portfolio of analytic capability with acquisitions such as SPSS and Cognos. They are a mainstay in the field and offer a variety of technical solutions.

When reading through the various summaries of advanced analytics, there are some obvious themes. By tearing apart each of these definitions, we will see that there are three fundamental components to a general definition of analytics. The first observation is that each of the four definitions eludes to some outcome or answer. By inference, the natural precedent for outcomes and answers is asking a question. Therein lies the first of the three fundamental components in

---

* http://www.miningyourownbusiness.com/wb/img/Elder_Research_eBook_The_Ten_Levels_of_Analytics.pdf/, Retrieved March 1, 2017.
† http://www-01.ibm.com/software/data/infosphere/what-is-advanced-analytics/, Retrieved April 15, 2015.

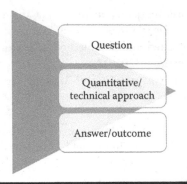

Figure 11.1   What is analytics?

analytics: a question. Analytics is an interrogative process wherein the practitioner is seeking to answer some question—more on that moving forward. With a question as the first fundamental component, the [previously referenced] answer or outcome is the second fundamental component. A key characteristic referenced in several of the definitions—also explained in more detail moving forward—is that the outcomes and answers in analytics are structurally related to, and derived from a set of underlying data. The first two parts being question and answer, the third fundamental component is that the outcome or answer must be derived from the available data sources, using some quantitatively technical process of inductive or deductive reasoning.

In summary, consider the following definition of analytics moving forward:

Analytics is an effective, comprehensive process of implementing a quantitative, technical, and data-driven approach to identify useful outcomes or answers to well-defined questions (Figure 11.1).

Importance of Analytics

Given the definition(s) of analytics given above, there is likely some natural intuition about why to undertake an analytics project, and such intuition is valuable. However, for the sake of clarity, it is worth fleshing out these concepts. To better understand, it might help to start with a more specific question, then scale back a bit. Here goes: Not just why do analytics, but why do analytics *now*?

The truth is that analytics has been around for a long time, but something has changed in the world, enabling a whole new host of analytic capabilities. Technology has gotten cheaper—significantly, exponentially, orders of magnitude cheaper. This has allowed everyone and everything to generate, collect, store, transport, and analyze data like never before. The facts and figures change too rapidly to have any permanence, but suffice it to say: available data has grown significantly, and the ability to analyze that data has grown significantly as well. This has resulted in a paradigm shift—a word often overused by consultants, but likely an understatement in this context—in the way questions are being asked, and outcomes and answers are being generated and delivered.

The result is that you should do analytics…because you can. There is more data available today than ever before. The quantitative, technical approaches that exist are more sophisticated, reliable, and efficient than ever before. Any analysis can be done faster, more accurately, and with more precision. Beyond simply coming up with outcomes and answers, results can be verified by subject matter experts, statistically validated, and returned to decision makers in shorter and shorter amounts of time. Most importantly, all of this can be done in a scalable, extensible, and repeatable way.

Part 2—How You Should Do Analytics (Process)

Introduction

It is not only important to understand what analytics is and why to do it, but also equally important is how to do it. While there is no singular prescription for how to successfully complete analytics projects, there is a wide body of experience that can help inform the process. Discussion of three major analytics process frameworks will help inform a more general analytics process. Some attention to the various steps of the generalized will help ensure your success.

Analytics Process Frameworks

In the same way, there is healthy discussion about how to define analytics, there also exists a good amount of discussion about *how* to do analytics. The single-most important aspect to completing analytics projects successfully is the fact that analytics is a process. Below there will be extensive discussion about various steps within an analytics process, but the key takeaway must be that each step is critically important to success. When conducting analytics projects, it becomes obvious that practitioners, stakeholders, and others engaged in the process will likely have an affinity for one step or another. In addition, a natural breadth of scope, sophistication, and scale exist for each step such that each step will not require the same attention for each project. For example, asking the right business question may not ever be as complicated as evaluating model performance. However, that simply does not mean that asking the right business question is less important. Each analytics project will require a different balance of skills and attention, and just as one practitioner may advocate the importance of one step on a project, there will undoubtedly exist a different practitioner on a different project that will make a compelling argument for an entirely different distribution of resources. This does not make one practitioner right and one wrong; it simply highlights the fact that each project offers its own challenges, its own benefits, and each practitioner may be in tune with the variability of demands.

Another problem understanding the effort and scope of the different steps in an analytics processes is the unequal attention these skills receive in education and training. Recent numbers suggest there are hundreds of educational programs in analytics. Many exist formally at the collegiate level, but the rapid growth of the analytics marketplace has put advanced analytical skills in high demand, creating an emergent need for less-formal training as well. Unconventional training curricula have appeared in the form of boot camps, shorter-form programs focusing on the fundamental quantitative and technical concepts of analytics. While the quality of such programs can be at a very high level, programs still make choices about how much emphasis to place on each step. This unavoidably sets a tone for how much time or effort each step will take in practice, and are thus unable to effectively convey the high variability that exists in practice. Each step in the analytics process is critical to success in analytics projects, but there is no general recipe that will work for every endeavor. Projects will require more attention during one step over another, based on unique characteristics of the project.

Borrowing from some of the sources used above, there are three analytic process frameworks useful in building out a generic process. Not surprisingly, there is tremendous overlap in each of the frameworks presented. But, there does not exist a simple one-to-one mapping. For this reason, each framework will be presented in summary, and then the generalized framework will be built out from there.

The SAS® Analytical Life Cycle

SAS takes an approach to analytics with an emphasis on *action*—they call it analytics in action (Figure 11.2). The process consists of three fundamental concepts: data, discovery, and deployment. As Figure 11.2 shows, discovery and deployment are comprised of a series of seven distinct steps, all of which are supported underneath by data. Across all the education and training referenced above, a tremendous amount of effort goes in to teaching analytic discovery. That makes a lot of sense, as a lot of the technical

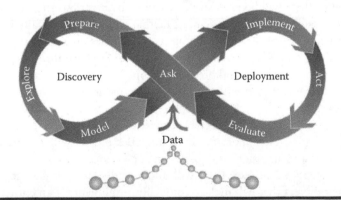

Figure 11.2 The SAS® Analytical Life Cycle: Best practices for improving your predictive modeling results.*

* https://www.sas.com/content/dam/SAS/en_us/doc/whitepaper1/manage-analytical-life-cycle-continuous-innovation-106179.pdf/, Retrieved March 1, 2017.

skills gap exists in that domain. And, to be clear, an analytic project will never be successful without a meaningful, data-driven insight—discovery is inextricable from analytics. However, the visualization depicting the SAS analytic framework highlights an earlier point that successful analytics projects require an analytics process comprised of something beyond analytic discovery.

The CRISP-DM Process: Cross-Industry Standard for Data Mining

A second popular methodology was developed in the mid-1990s. Commonly known by the acronym CRISP-DM, the Cross-Industry Standard for Data Mining has emerged as a popular method for engaging in analytics projects, not just data mining specifically. Originally developed by SPSS, Inc. (a software company later acquired by IBM), the CRISP-DM process places significant emphasis on business understanding and data understanding. Figure 11.3 illustrates the iterative, back-and-forth process

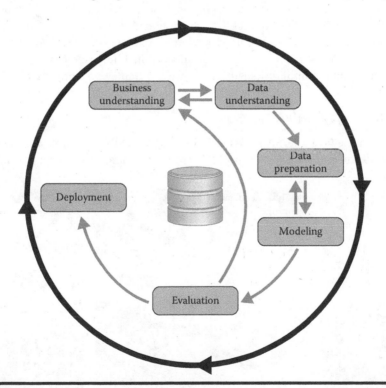

Figure 11.3 CRISP-DM process diagram showing the relationship between its different phases.*

---

* ftp://public.dhe.ibm.com/software/analytics/spss/documentation/modeler/18.0/en/ ModelerCRISPDM.pdf/, Retrieved March 1, 2017.

that exists between business and data understanding, as well as the use of evaluation output to feed back into the business understanding. The use of these feedback loops is an important part of each analytics process.

INFORMS Certified Analytics Professional, Job Task Analysis (JTA)

A third framework for consideration was developed by the Institute for Operations Research and Management Science (INFORMS). In 2010, INFORMS partnered with Capgemini to evaluate the analytics marketplace, and their potential role within it. One of the findings from that study indicated there was a need for analytics certification. After assembling a team of experts in the field; from academia and practice, private and public sector, multiple industries, and multiple fields of practice; INFORMS created the Certified Analytics Professional (CAP) designation. A part of this development process was identifying necessary skills an analytics practitioner must possess to ensure success undertaking analytics projects. INFORMS developed the Job Task Analysis (JTA) to highlight the different phases of an analytics project, eventually becoming their steps in an analytics life cycle (Figure 11.4). Beginning with Business Problem Framing and culminating in Life Cycle Management, the process shares many similarities with the other two frameworks presented above.

Step-by-Step Analytics

A careful review of the SAS, INFORMS, and CRISP-DM frameworks reveals subtle difference to each of the approaches. Despite these nuances, one thing that

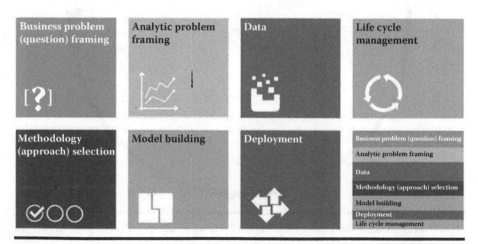

Figure 11.4   INFORMS Certified Analytics Professional: Job Task Analysis (JTA).*

* https://www.certifiedanalytics.org/download.php?file=caphandbook/, Retrieved March 1, 2017.

SAS Analytical Life Cycle	INFORMS JTAs	CRISP-DM
1. Ask	Business problem framing	Business understanding
		2. Data understanding
	3. Analytics problem framing	
4. Prepare	Data	Data preparation
5. Explore		
	6. Methodology	
7. Model	model building	Modeling
		8. Evaluation
9. Implement	Deployment	deployment
10. Act		
11. Evaluate		
	12. Model Life cycle Mgmt	

Figure 11.5   A dissection of the three frameworks.

stands out is a co-linearity that exists between each framework. That is, all three approaches approximately begin in the same place, end in the same place, and follow the same general trajectory along the way. Integrating these three frameworks into one singular step-by-step process will help flesh out a more generalized framework. Employing such a framework in analytics projects will help maximize the likelihood of success. While each of the following concepts originates from one of the three frameworks, each will be discussed on its own merits and will not be tied to a source (Figure 11.5).

Business Problem Framing

Fundamentally, analytics problems are business problems. Regardless of your organization's objective, be it finance, retail, national security, community service, or something altogether different from anything else, at its core, there is some purpose and objective to its existence. For the purposes of analytics, the business problem is the problem you want to address. The first step in an analytics project, is to fully understand what that problem is. It is all too common to begin an analytics project and realize at some point further into the project that a tremendous amount of effort was being expended to solve the wrong problem. Therefore, putting in more up-front work to determine what that problem is and very narrowly define its unique characteristics, is critically important.

There are several popular ways to accomplish the goal. One common process is to undertake a five why root cause analysis. Many times, the first pass at a business problem fails to truly address the underlying issues. For example, if your organization wants to improve customer retention, a five why analysis will encourage you to go a step further: why do you want to improve customer retention? "Well," you might say, "that's obvious. We're losing a ton of money on customers that leave and go to another provider." The five why analysis will refocus your problem to "losing a ton of money." So, what initially sounded like a customer retention issue may be a financial problem. Is the root cause of the financial issue that customers are leaving? Or is it that the wrong customers are leaving? Perhaps the high-margin customers are leaving, but the low-margin customers are staying. If high-margin customers are leaving, and low-margin customers are staying, it may be more than just "customers are leaving" and it may be a broader revenue management issue. Following this same process, you may say "why is there a revenue management issue?" You may find that there is an outdated enterprise revenue management solution. "Why is the revenue management solution outdated?" Perhaps the current enterprise data management solution does not offer revenue management, or maybe the capability they offer requires special IT systems that your organization does not currently possess. If some of that nuance is unclear, you may need to abstract the problem to your own business scenario, but in this exercise, what began as a customer retention issue may be an IT issue. Viewing the problem through this slightly different lens may open minds to the real problem, and that is where you want to focus your analytics resources. For the sake of thoroughness, in the example discussed above, you might consider investing in an analytic solution to help optimize pricing and revenue management.

Data Understanding

Since there are several different ways to think about data, Data Understanding can mean a lot of different things to different people. The IT manager may think about Data Understanding in terms of what type of data storage solution is in place? Is it a relational database, is it column store, graph, or key store? A database administrator may think more about the characteristics of its structure. How many columns are there? How many records? Is there an effective method of reduplicating, such as primary keys? A data scientist may be interested in completeness of data, whether it is continuous or discrete, or whether any entity resolution has taken place. A final consideration may be the stakeholder themselves. Where is the data coming from? Are the sources reliable? Are these the only data available? Are they the only data necessary?

Simply put, these questions are important in the success of an analytics project. Each role described above plays a vital part in making the analytics project successful. It is critical to have a complete and comprehensive understanding of the data, where they are from, how they are stored, and what information is contained in those data. In addition to understanding these data, documenting this information along the way can help avoid confusion later in the process. Understanding which data sources are available, how they are being used, where they are from, and any known issues will help in subsequent processes, such as model building and deployment.

Analytics Problem Framing

In addition to understanding the business problem, it is important to translate that business problem into an analytics problem. The definitions of analytics discussed above each suggest that analytics is not just one tool, but a variety of tools. One analysis by Andy Fast and John Elder of Elder Research, identify ten different levels of analytics.* Beginning with Standard and Ad Hoc Reporting, continuing to statistical analysis and learning, optimization, ensembles, and causal modeling (not exhaustive), they are very intentional about what they include on their list of analytics, and it does not exclude some of the more basic analyses. Each discrete level of analytics they mention may have a nearly boundless list of tools and capabilities with which it is associated. To be successful in an analytics project, it is important to identify from the innumerable which specific techniques will help solve the business problem. This is the process of analytics problem framing—how do you translate the business problem you are addressing into a problem that can be solved by one or more analytics solutions.

Data Preparation

Much of the data used in analytics projects are what data scientists refer to as found data. Except in rare cases, data are almost always collected for a different reason than your analytics project. Consider payroll data, academic transcripts, voltage or current readings on a power transmission line. While this data may be useful during your analytics project to help you answer your question or address your business problem, the truth is that these data were collected for their own reasons, and the analytics practitioner is merely repurposing them for her purposes. With that in mind, it is very common to find data in a less than desirable format for conducting analytics. Some data is stored as the wrong type (e.g., string versus numeric), other data are incomplete (e.g., optional data fields, like birthday), other data are stored in the wrong storage architecture (e.g., social media networks may be in a graph database). Part of the analytics process is cleaning, organizing, and preparing the data to be used for an analytics project. Certain analytical techniques may require something like an analytic base table (ABT), a very wide data format where each record represents a unique entity and every possible predictor and target variable for each entity are listed as columns. Preparing data may also involve moving and merging data to get them into the best place

---

* http://www.miningyourownbusiness.com/wb/img/Elder_Research_eBook_The_Ten_Levels_of_Analytics.pdf/, Retrieved, Retrieved March 1, 2017.

for optimal analysis. Transporting large data sets across a network, or performing complicated data merges (i.e., bringing two or more tables together as one), may be network or computationally intensive. Putting data in the right location and performing basic data processing tasks, so-called preprocessing, can dramatically improve the time it takes to do analytics.

Explore

One of the characteristics frequently cited as common among talented analytics practitioners is their common sense of curiosity. In many ways, having a curious mind is more of a personality characteristic, and less of a best practice. However, it is always important to *play* in the data, a term commonly used to describe the exploratory process of interacting with data sets. There are a variety of ways to explore data sets, but many involve some type of light prototyping.

For example, a first look at a new data set may involve running various statistical tests to look for any obvious correlations. Summary statistics, including information about the distribution of the data may give insight into general trends or patterns of interest. Analytics practitioners may also choose to run data through one or more visualization capabilities to see if there are any readily observable characteristics (e.g., are data clustered tightly, or more uniformly distributed). Doing a first pass with the naked eye can help direct you to a set of analytics that may be particularly useful for the type and characteristics of the data in mind. Similarly, it may be helpful to create some rudimentary statistical models to determine whether any patterns emerge. By running a few simple algorithms, you can start to determine whether there are any natural patterns. As much as this may help find useful ways to analyze the data, it may just as easily find techniques that will not be useful. Data analysis in this process can be thought more in terms of divergent thinking, rather than in convergent thinking.

Methodology

Once this data has been explored, it is time to determine which specific methodologies will be employed. Unlike before, where there was some light prototyping and exploratory research, here the goal is to engage in more convergent thinking, rather than divergent. After trying out various analytics approaches, this is the part of the process where you start to focus on which specific analytics methodology will help solve the business problem.

A cautionary note about methodology or model selection: it can be tempting to use the most advanced, sophisticated approach available. There is a false equivalency between complexity and performance, and, in fact, in many situations the opposite is true. Complex analytic solutions certainly have their place, but frequently what accuracy or precision in output you gain with complexity, you lose intuition and ability to clarify. There are many good cases of technical success, but with business failure. The reasons for this are myriad, but many times it simply revolves around the stakeholder's ability to implement the change. More complex models are inherently more difficult for stewards of company resources to understand, and will adversely affect likelihood of analytic adoption.

Model Building

There are several different ways to build the models necessary to solve your analytic problem. There is no shortage of opinion about which tools are best for various analytic techniques. In truth, many of these are overstated, as a good practitioner can generally accomplish the task with more than one tool. True agnostics are equally overstating their case. The truth lies somewhere in between.

A frequent conversation had in analytics shops throughout the marketplace is whether to choose a licensed (read: expensive) or open-source (free). The debate about these two paradigms is not quite as straightforward as both solutions can be part of a viable analytics strategy. For many analytics projects, an open-source tool can facilitate rapid prototype development. Work completed in the exploration stage discussed above is ripe for open-source technologies. A capable practitioner can download an open-source analytics software package, connect to some data, and begin building a modeling solution in as little as a few hours! But, as discussed in several concepts above (e.g., SAS Analytics in Action), and will be discussed further below, model development is on one piece of the puzzle. Exclusive focus on model building neglects major concerns related to productionalization.

There is another important discussion point about licensed versus open-source software. It is a common mistake to confuse purchase price with total cost. Clearly, going to a website and downloading the latest version of Python may, in fact, have zero monetary cost. However, this distorts an all-too-common reality that using open-source software tools frequently takes much more time to develop. Further, the resources used to develop with those open-source analytic tools are often paid a premium for their skills. Simply put, licensed software is designed to be user-friendly, and more approachable. A separate problem affecting the open-source ecosystem is their lack of enterprise support and deployment capability. Frequently, the money saved on licensing fees for analytics software are offset by the added costs of development time, developer compensation, and what can be significant additional resources to deploy the solution. This will be discussed again below, but it is worth mentioning here when you make decisions about which technology you use to build the analytics solution.

Evaluation

There are three important concepts in evaluation: verification, validation, and model performance. Verification is the technical process to determine whether the model you built does what you designed it to do. Validation is a process used to determine whether you have chosen the correct modeling technique. Last, model performance is a set of technical approaches to determine how well your model performs.

Model Deployment

Model deployment is the stage of this process that receives the least amount of discussion in current literature. Without deployment, even the most successful analytics solution to your business scenario is not creating an impact. Creating the modeling approach and solving the hard problems can only contribute to your organization's performance when you use the model to help make a business decision. For this reason, it is critical to have a deployment strategy from the beginning. Before pulling together data, going through modeling approaches, and building a model, it is imperative that you think through how the analytic result will be used. You must use an analytic solution that you can deploy into your operational environment. In many scenarios, that means deploying your analytic solution within some type of IT infrastructure. For example, if you build a credit scoring model, you need to have those results available to the enterprise system when a user submits their application for credit. That process can be automated, or built in to the logic of the application and approval system, or it can be manual, where the results of the analytic output may be available to a loan officer or underwriter. Regardless how you serve up this analytic result, the reality is that unless you have those results available for use during the business process, the analytic will not be able to have an impact on your business problem. Another common use of analytics is to provide the insight back to decision makers, either through a dashboard or report of some type. Again, the best, most interesting insight will be useless if a decision maker does not have access to the information at the time they need to make the decision. This highlights again the need to consider deployment when building a model. While open-source analytics solutions appear inexpensive because their free price tag, deploying those open-source analytic results into your business process may prove much more challenging than appears at first glance. Deployment is a key differentiator for licensed software solutions, as many good analytics software platforms have given deployment the right amount of attention.

Act

Once the modeling solution is built and deployed, the business needs to act on that information in real time. At the point of decision, the business process needs to act on the analytics solution on an ongoing basis. It is not uncommon to see analytics projects find and build the right solution, deploy that solution, and then at some point in stop working. Key business processes may be reverted to an old way of doing business without the analytics practitioner ever knowing. Without continued attention and vigilance, analytics may be built, they may be deployed, but some business process may change external to the analytics solution and the organization is no longer acting on that information it needs to.

Evaluate

An important part of the analytics process is to conduct periodic reviews of the continued validity of the analytics solution. By evaluating the business life cycle of an analytic solution, you are constantly reassessing to determine if circumstance have changed to warrant changes or modifications to your modeling approach or analytics solution. It is surprising to see how frequently a business process changes rendering analytics result less relevant. Many times, the analytics solution will remain in deployment, decisions continue to be made based on the analytic insight, but the analytic is no longer answering the right business question. In these circumstances, your analytics solution may be working against you. If this happens, it is time to go back to the earlier steps to re-engage with your business problem and reframe your analytics solution.

Model Life Cycle Management

Separate from the business life cycle of analytics solutions, analytic models have their own life cycle. For example, predictive models are used to help identify whether some event will happen in the future. Ostensibly, successful predictions of those outcomes will give your organization the ability to change its behavior—to make more or less of something happening. By changing the business process to effectuate the change you want, you can materially change the environment about which you are predicting. An example of this is when you take some action to improve cross-selling performance. Supposing you run some analytics to build a recommendation engine for your website. When a buyer is interested in buying product A, you mention to them that buyers who historically have bought product A have frequently also bought product B. If after a year of making this recommendation, you continue to recommend product B every time someone buys product A, you may fail to account for the fact that users are buying product A and B at the same time because you are recommending it to them. While this is not necessarily a bad thing, you would still want to know whether their interest in product B was of their own choosing or was based on your recommendation. When you use models to make business decisions, there is a life cycle need to continuously re-evaluate the business scenario at hand. Similarly, by not properly considering life cycle management of your analytics solutions, you may not catch changes in the underlying patterns of behavior. One common description of this is not being aware of "unknown unknowns." If you are trying to predict power consumption over the next several years, and you base these predictions on historical data, if a major change in technology occurs, you will need to re-evaluate your modeling approach to account for the new technological change.

Conclusion

This chapter provides a foundation for understanding analytics, how to use them, what analytics processes and frameworks can be used, and better comprehending the steps in the analytics process. Analytics, coupled with intuition based on experiential learning, is a powerful combination for management decision-making. The years ahead look bright for this emerging field.

Chapter 12

Data, Insights, Models, and Decisions: Machine Learning in Context

Stephen Adams, William T. Scherer, and
Peter A. Beling

Contents

Machine learning is a collection of techniques designed to detect hidden patterns in data and construct or learn models for predicting the outcome of future data. These predictions can be used to make decisions about the system generating the data in the face of uncertainty. In the traditional paradigm, the work flow is data to model to decisions, and the decisions are essentially solely influenced by the data. This type of purely data-driven prediction system relies on the assumption that all properties of the system are captured by the collected data. In some applications, this assumption can hold. However, when data is sparse, when it is difficult to collect

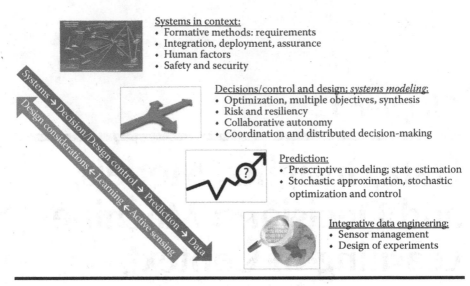

Figure 12.1 Systems in context.

an adequate data set, or when the data is too complex to model, the traditional machine-learning paradigm can produce suboptimal results.

In this chapter, we will propose adding insights, intuition, or prior knowledge about a system into the modeling process. Figure 12.1 illustrates our philosophy regarding the proper approach to predictive modeling and other forms of machine learning. The fundamental idea is that machine learning must be done in the context of the problem being explored, and this inherently mandates the inclusion of this additional knowledge that traditionally is not included in machine-learning discussions. This "contextual integrity" provides a framework for solving a predictive analysis with all the appropriate knowledge included (Gibson, Scherer, Gibson, & Smith, 2016).

Contextual understanding and insight into systems can come in many forms and from numerous sources. Generally, insight about a system is provided by domain experts or from a system analysis. A practitioner wishes to incorporate the prior knowledge about a system into the model so that it captures physical properties of the system and/or constraints of the system. Understanding of system properties and constraints are combined with an understanding of the design or operation decisions to be made to formulate a computational decision problem. It is only at the point when a decision problem has been formulated, that machine learning is brought to bear on the available data. We assert that the insight acquired during system modeling and problem formulation should be expressed in the form of prior probabilities on the parameters of interest in the decision problem. Using this type of information will produce models that more closely represent the system being studied. An analyst may know, for example, that a Markov transition matrix has a special structure, such as lower-diagonal, and can then use this knowledge to better structure the machine-learning process. A naïve learning algorithm may not recognize such a structure and produce incorrect and/or inefficient predictive models.

Prior knowledge can also influence the modeling process. Machine-learning models can often make assumptions about the process or the data. For example, some models assume that the data follow a Gaussian distribution. If this assumption does not hold, the learned model can underperform or completely breakdown, providing erroneous predictions leading to poor decisions. Alternatively, a state-based model's (dynamic or discrete) initial conditions, generated by human experts, may be critical to the transient response of predictive machine-learning models. An understanding of the system, therefore, is crucial before modeling can begin. Different assumptions can lead to wildly different outcomes in terms of model selection or accuracy. Generally, it is best to back up an assumption with insights or details about the system or process being modeled.

We will discuss a mathematical formulation for incorporating insights, intuition, or prior knowledge into the modeling process using informative prior distributions. Then, we use this framework to consider three cases—from the disparate domains of medicine, manufacturing, and consumer credit—that share the characteristic that prior information is both knowable and useable.

Mathematical Framework for Intuition

One area of machine learning is focused on learning parametric models that explain the behavior of the data. These types of parametric models often make assumptions about how the data interact with an outcome or the distributions generating the data. We propose Bayesian statistical analysis as one method for incorporating intuition and prior knowledge into parametric models.

Bayesian methods rely on Bayes Theorem that relates the conditional probability of events given some idea about how likely that event is to occur. We will describe Bayes Theorem using an example with two events A and B. Let the probability that event A will occur be $P(A)$, and likewise let $P(B)$ be the probability event B will occur. These probabilities can be referred to as marginal probabilities, prior probabilities, or unconditional probabilities. A conditional probability is the probability that an event will occur, given that another event has occurred. In our simple example, one conditional probability is the probability that A will occur given that B has already happened, and this probability is denoted as $P(A\,|\,B)$. The conditional probability of A given B is related to the conditional probability of B given A through Bayes theorem

$$P(A\,|\,B) = \frac{P(B\,|\,A)P(A)}{P(B)}$$

In Bayesian statistics, all parameters are treated as random variables. In a general setting, let θ represent the set of model parameters and let X represent the collected data. A purely data-driven model attempts to learn parameter estimates ($\hat{\theta}$) by maximizing the probability of observing the data given a set of model parameters

$$\hat{\theta} = \underset{\theta}{\operatorname{argmax}}\ P(X \mid \theta)$$

$P(X \mid \theta)$ is often referred to as the likelihood of the observed data given the parameters θ, and this method is called maximum likelihood estimation. Bayesian learning, on the other hand, places a prior distribution on the model parameters and maximizes the posterior distribution $P(\theta \mid X)$. This method is called maximum a posterior (MAP) estimation. Using Bayes rule, the parameter estimation problem becomes

$$\hat{\theta} = \underset{\theta}{\operatorname{argmax}}\ P\big(\theta \mid X\big) = \underset{\theta}{\operatorname{argmax}}\ P(X \mid \theta) * P(\theta)$$

where $P(\theta)$ is the prior distribution. Note that $P(X)$ is excluded from MAP estimation because this probability is fixed and does not influence the parameter estimates.

In Bayesian estimation, the prior distribution on the model parameters is typically noninformative, meaning that it assigns equal weight to all possibilities. Priors are chosen by the practitioners, researchers, or modelers; therefore, the practitioners can influence the estimation by the selection of a prior distribution or the parameters for that distribution. Noninformative priors are used when there is a desire for the data to be the only factor driving estimation, and these priors prevent any bias or influence being injected into the estimation by the practitioner. Furthermore, noninformative priors allow for easy selection of prior distributions and hyperparameters, the parameters of the prior distribution, as these choices have little effect on the posterior so long as they remain noninformative. Note that MAP estimation using noninformative priors is equivalent to maximum likelihood estimation.

The cartoon in Figure 12.2 illustrates a principal aspect of the Bayesian perspective. The frequentist statistician knows that the probability of the detector lying is low (1/36) and therefore concludes that when the detector says the Sun has exploded then it is likely to have done so. The Bayesian statistician knows the prior that the Sun has exploded in each day is 10^{-13} or less (our Sun won't explode but will instead become a red giant star) and therefore concludes it is exceedingly likely that the detector is lying, as the posterior probability of the Sun having exploded given the detector result is still around 10^{-13}.

On the other hand, informative priors can be used to convey intuition about a system to the estimation process that is not readily available in the collected data.

Informative priors are often avoided for two reasons in Bayesian analysis. The first is the previously discussed researcher influence that can affect the analysis. The second is the lack of a consistent methodology for constructing informative priors that is widely accepted over types of models and fields of study. It is reasonable to assume that two different researchers could construct very different prior distributions, given the same prior data or set of assumptions.

Figure 12.2   A Bayesian makes a bet.*

* https://xkcd.com/1132/

Informative Priors

In this section, we give a brief overview of the use of informative priors in research and literature. Several of these ideas will be discussed in detail in the following examples.

The use of informative versus noninformative priors has been widely debated. Jaynes (1985), who outlines the historical debate up to 1985 and favors the use of informative priors, presents the following example of how prior knowledge is used in every-day inference: A medical diagnostician would not make a diagnosis using only the patient's current condition or symptoms and ignore the patient's medical history. Jaynes concludes by showing how the use of highly informative priors on seasonal parameters for monthly economic data improves forecasting.

Thomas, Witte, and Greenland (2007) argue for the use of informative priors in epidemiological studies using hierarchical models in their paper, which is subtitled "Who's afraid of informative priors?" They state that letting the data "speak for themselves" assumes that the collected data are a good representation of the system being modeled and correct. Further, they state that data by itself is not useful without assumptions about the model and the inclusion of prior information. They assert the need to balance data with prior assumptions and information, which directly aligns with our argument for the need to model systems with intuition and in context. Zero-numerator problems involve probabilities, such as a transition matrix for a Markov model, that do not occur in the training data but are possible. If using a maximum likelihood estimator during training, their exclusion from the training data causes the probability estimate to be 0, even though the true probability is greater than 0. Winkler, Smith, and Fryback (2002) study the role of informative priors in zero-numerator problems, where an event is possible but the probability of the event occurring could be estimated to be zero if the event is not present in the collected data.

Similarly, Lord and Miranda-Moreno (2008) demonstrate that low sample means for Poisson-gamma models and small sample sizes can affect posterior distributions when vague or noninformative priors are used. Using a nonvague prior decreases the chances of poor parameter estimation. Jang, Lee, and Kim (2010) propose a power prior for zero-inflated regression problems.

Informative priors have been used in several types of models. Angelopoulus and Cussens (2005) use informative priors in classification and regression trees. Coleman and Block (2006) use informative priors in the estimation of parameters for nonlinear systems. Mukherjee and Speed (2008) study informative priors in network inference.

While previous studies have demonstrated that good informative priors can improve parameter estimation and the predictive ability of models, there is no well-established methodology for converting prior knowledge into prior distributions. Guikema (2007) compares five methods for constructing prior distributions for failure probability estimation in reliability analysis: the method of moments, maximum likelihood estimation, maximum entropy estimation, noninformative prepriors, and confidence/credible interval matching. He demonstrates that the assumptions on the prior greatly affect the posterior. He concludes that if the data used in constructing the prior accurately reflects the data collected later, maximum likelihood estimation gives a posterior with the minimum variance, but maximum entropy estimation is the most robust to differences between the prior data and the observed data. In another study, Yu and Abdel-Aty (2013) compare four methods for constructing informative priors: two-stage Bayesian updating, maximum likelihood estimation, the method of moments, and expert experience. Each of these has strengths and weaknesses, but the authors conclude that the two-stage Bayesian updating procedure is superior for the data and model used by the authors.

Raina, Ng, and Koller (2006, June) propose the use of transfer learning for constructing informative priors for text classification. Vanpaemel (2011) uses hierarchical methods for constructing informative priors, and Washington and Oh (2006) outline a methodology for building informative priors from experts' opinions on ranking railroad countermeasures.

Rare Disease Priors Example

Microcephaly is a rare birth defect where the brain develops abnormally. Depending on the source, the probability of being born with microcephaly can range from 12 in every 10,000 live births* to 1 in every 666,666 individuals in the United States†. If the mother contracts the Zika virus in the first trimester of a pregnancy, the risk that the child is born with microcephaly can increase to between 1% and 13%‡. In this example, we will demonstrate how insight and prior knowledge can be harnessed to estimate the probability of being born with a rare disease.

For each person, there are only two possible outcomes: they are either born with the defect or without the defect. Let Y be the random variable indicating the presence of the birth defect, where $Y = 1$ if a person is born with microcephaly and $Y = 0$ if they are born without the defect. For any individual, let θ be the true but unknown probability that $Y = 1$

$$\theta = P(Y = 1)$$

Our goal is to find a parameter estimate for θ; we use $\hat{\theta}$ to indicate our estimate of θ. A practical use for this estimate is predicting the number of births with the defect in each year or within a given geographic location. This prediction could affect

---

* https://www.cdc.gov/ncbddd/birthdefects/microcephaly.html
† http://www.toptenz.net/top-10-rarest-diseases.php
‡ https://www.statnews.com/2016/05/25/zika-microcephaly-what-is-risk/

long-term decisions such as monitoring of at risk births or providing funding for care. We can model the number of births with microcephaly, given the probability θ and the total number of births N, using a binomial distribution. If $x = \sum_{n=1}^{N} Y_n$, then

$$P(x \mid \theta, N) = \binom{N}{x} \theta^x (1 - \theta)^{N-x}$$

To use this model for prediction and decisions, we first need the estimate of θ. One way to estimate this quantity is to collect data a sample from the birthing population. In the binomial model, the sample size will equate to N and the number of births with microcephaly in the sample will be X. The maximum likelihood estimate for the probability is

$$\hat{\theta}^{ML} = \frac{x}{N}$$

which is the sample proportion. Often this estimate is sufficient; however, when the $P(Y = 1)$ is small, there is a chance that the sample could contain no instances of microcephaly. Using the maximum likelihood estimate would yield $\hat{\theta}^{ML} = 0$. We know this is not the case because there are in fact instances of births with microcephaly. To overcome this zero-numerator problem, we will use the Bayesian technique to estimate the probability.

For a binomial distribution, the conjugate prior for the proportion parameter is a beta distribution

$$P(\theta \mid a, b) = \frac{\Gamma(a+b)}{\Gamma(a)\Gamma(b)} \theta^{a-1} (1 - \theta)^{b-1}, \quad 0 \le \theta \le 1$$

where Γ is the gamma function. The posterior distribution of θ is

$$P(\theta \mid x, N, a, b) = \frac{\Gamma(N+a+b)}{\Gamma(x+a)\Gamma(N-x+b)} \theta^{x+a-1} (1 - \theta)^{N-x+b-1}$$

which is a beta distribution with parameters $x + a$ and $N - x + b$. The maximum a posteriori estimate for the probability of being born with microcephaly is the mean of this beta distribution

$$\hat{\theta}^{MAP} = \frac{x+a}{x+a+N-x+b} = \frac{x+a}{N+a+b}$$

Even if there are no instances of microcephaly in our sample, the MAP estimate for θ will always be greater than 0 because both hyperparameters in the beta prior must be greater than 0.

Selecting the hyperparameters in the prior distribution can be difficult, and this is where insight into the system plays a large role. If very little information is known about the system, a uniform prior is often the optimal choice in Bayesian estimation. A uniform prior for this example puts equal weight on all value between 0 and 1. For a binomial distribution, a uniform beta prior has $a = b = 1$. If we have a strong belief that θ is small, we can set $a \ll b$. However, selecting how much less a should be than b is also quite difficult.

In our microcephaly example, we have prior studies that suggest θ is somewhere between 1.5×10^{-6} and 0.0012. We can use this information to assist in selecting hyperparameters. Let θ_0 be our prior belief about the value of θ, and let σ_0 be our uncertainty in this belief. We can set our prior belief to the mean of the beta prior

$$\theta_0 = \frac{a}{a+b}$$

and our uncertainty in this belief to the standard deviation of the prior

$$\sigma_0 = \sqrt{\frac{ab}{(a+b)^2(a+b+1)}} = \sqrt{\frac{\theta_0(1-\theta_0)}{a+b+1}}$$

Given our belief and uncertainty in our belief, this system of equations can be solved to give us the hyperparameters in prior distribution. Let $\theta_0 = 0.000601$ (the mean of the two extremes form the prior studies) and $\sigma_0 = 0.01$ (a large amount of uncertainty given the low proportion). This approximately yields $a = 0.003$ and $b = 5$. Using these hyperparameters, if a sample of 10,000 births does not contain a single instance of microcephaly, then $\hat{\theta}^{MAP} = 3 \times 10^{-7}$. This is not within the range of the previous studies, but the estimate is nonzero. This is significant because a zero probability implies that something is impossible and there is clear evidence that cases of microcephaly exist.

FSHMM Example

Feature selection is the process of selecting a subset of relevant features from a larger set of collected features (Blum & Langley, 1997; Dash & Liu, 1997; Guyon & Elissee, 2003). Feature selection can improve the predictive ability of models by removing redundant and unnecessary noisy features. By selecting a subset of data streams instead of the set of all possible data streams, the required data storage and processing times of algorithms is reduced. Furthermore, feature selection can help with reducing the cost of system design by limiting the number and types of sensors that must be used to collect data. Generally, feature selection methods evaluate a feature subset using some measure of predictive accuracy and some notion of model complexity.

These general feature selection techniques unknowingly assume that all data streams have the same cost. Cost is not limited to financial cost and, depending on the application, could refer to the difficulty in acquiring the data, time to acquire the data, or the accuracy of the data. In real-world applications, data streams can have wildly varying costs. Consider the problem of predicting cutting tool wear in rotary machines such as CNC lathes. Tool wear cannot be directly assessed during the cutting process without stopping the process, which would cause the company to incur production delays and, in some cases, degrade the final quality of the product. Sensors can be installed on the rotary machines to collect data, and the collected data can be used to train machine-learning algorithms for predicting tool wear. There are several possible sensors that can be used to collect the data including force, vibration, and acoustic emissions. There can be significant differences in the price of these types of sensors. Force sensors can be on the order of $50,000, while vibration sensors are closer to $1,000. A general feature selection technique would not consider the difference in price when selecting a relevant feature subset, but the difference in price is certainly a concern when constructing a tool life prediction system for commercial use.

Informative priors can be used to incorporate cost into the feature selection process. Adams, Beling, and Cogill (2016) developed a feature selection method specific to hidden Markov models (HMMs) and hidden semi-Markov models (HSMMs) that considers cost. HMMs (Rabiner, 1989) are probabilistic models for time series data with two random variables. The state variable is modeled as a Markov chain and is usually unobservable. At each time instance, the state is accompanied by an emission that is observable. The emission random variable has a distribution that is dependent on the state variable. HSMMs (Yu, 2010) are extensions of HMMs that assume a duration or sojourn time for each state.

The feature saliency HMM (FSHMM) proposed by Adams et al. assumes that the emissions, represented by Y, come from a mixture distribution that is dependent on the state, represented by X. The mixture distribution is composed of a state-dependent Gaussian distribution and a state-independent Gaussian distribution, and it assumes that each of the L features is independent

$$P\left(y_t \mid x_t = i\right) = \prod_{l=1}^{L} \left(\rho_l * r\left(y_{lt} \mid \mu_i, \sigma_i^2\right) + \left(1 - \rho_l\right) * q\left(y_{lt} \mid \varepsilon, \tau^2\right)\right)$$

where y_{lt} represents the observation of the lth feature at time t, and $x_t = i$ indicates that the hidden Markov chain is in state i. The state-dependent Gaussian, represented by $r\left(y_{lt} \mid \mu_i, \sigma_i^2\right)$, has a mean of μ_i and a variance of σ_i^2, both of which are dependent on the state variable. The state-independent Gaussian, represented by $q(y_{lt} \mid \varepsilon, \tau^2)$, has a mean ε and variance σ^2, which are both independent of the state variable. The feature saliency ρ_l is the mixture weight and represents the probability an observation from the lth features belongs to the state-dependent distribution. The idea behind FSHMM is that relevant features should be more likely to come from the state-dependent distribution, and thus ρ_l can be interpreted as the probability that the lth feature is relevant.

In FSHMM, all model parameters are treated as a random variable. Priors are placed on each parameter and the MAP estimates are calculated using the expectation-maximization algorithm (Gauvain & Lee, 1994). Cost is conveyed to the feature selection process through an informative prior on the feature saliencies. A truncated exponential prior with support on [0, 1] is used

$$ P(\rho_l) = \frac{1}{Z} e^{-k_l \rho_l} $$

where Z is the normalization constant and k is the rate parameter. A large value for the rate parameter pushes the prior distribution toward 0 and thus conveys that more evidence is needed from the data to conclude that the feature is relevant. The cost of each feature is used to select the rate parameter where features with higher costs are assigned higher rates.

Figure 12.3 displays three standard exponential distributions with rates equal to 5, 10, and 25. Note that these plots are standard exponentials and not the truncated exponentials used in the FSHMM formulation. As the rate increases, most of the density is pushed toward zero. The expected value of an exponential distribution is the inverse of the rate parameter. Therefore, as the rate parameter increases, the expected value decreases. The expectations of the three plotted distributions are 0.2, 0.1, and 0.04.

To demonstrate this method, FSHMM is applied to tool wear data (Adams et al., 2016). Six independent sensors are used to collect data: force in three directions and vibration in three directions. It is assumed that each force sensor costs $2400 and that each vibration sensor costs $1200. Root mean squared error, the sum of log energies, and maximum energy are extracted from each sensor's data stream and used as features. Half of the assumed cost of each sensor is used as the rate parameter for each feature saliency prior distribution.

Several numerical experiments are conducted using this example. The MAP formulation of FSHMM is contrasted with a maximum likelihood formulation, which does not use priors and does not consider the cost of each sensor when selecting features. Full models that use all 18 features are compared with reduced models where three features corresponding to a single sensor are removed from the feature set. These experiments demonstrate that the accuracy of the reduced models is not

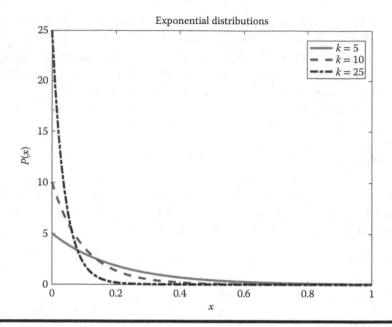

Figure 12.3 Plot of standard exponential distributions with rates equal to 5, 10, and 25.

significantly different from the accuracy of the full model, indicating that the feature set can be reduced. Further, the accuracy of the ML formulation is not significantly different from that of the MAP formulation. However, MAP outperforms ML in terms of cost. The ML formulation routinely removes a vibration sensor, while the MAP formulation removes a force sensor. Given that there is no significant difference between the accuracy of the models, the model with lowest cost should be preferred. The total cost of sensors using the results from the MAP formulation is $8400 while the total cost of the sensors using the ML formulation is $9600.

This example demonstrates how prior information that is not readily available in the collected data can and should be incorporated into the design of a system. Without this insight, an accurate and possible sufficient system could be designed. However, incorporating the prior knowledge about cost leads to a more profitable design.

Markov Model Example

In this example, we deviate from the informative prior framework for incorporating insight into modeling. We address how insight can be applied to the structure of a model, and how it can drive model building.

Data-driven modeling has become very popular in the last decade. Neural networks and deep-learning techniques have been shown to be extremely powerful

Inputs   "Black box"   Outputs

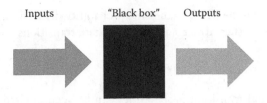

Figure 12.4 Black box model.

in problems such as supervised classification of images. We often think of these types of models as "black boxes" because their inner workings are not interpretable, and they can be characterized strictly as inputs, a black box, and outputs (see Figure 12.4). Neither the number of layers in a deep neural network nor the number of nodes in each layer provide insight into the problem or system being modeled. In some part, the popularity of neural networks and deep learning can be attributed to their black-box nature. Users do not necessarily have to have a deep theoretical understanding of the model, and software packages such as Tensor Flow allow for users to implement deep learning with almost no theoretical knowledge.

However, some view their black box nature as a drawback. Companies or sponsors can be reticent to utilize purely data-driven models without some link to the physical nature of the problem. Further, without the interpretability of the inner workings of these models, it can often be difficult to understand why a model predicted one outcome over another. In some fields, this interpretability is crucial to decision-making. In credit scoring, there needs to be justification for why a loan is denied. Typically, logistic regression models are used to predict the probability of defaulting on a loan because the model allows for the outcome to be traced back to a single or set of explanatory variables that most influenced the prediction. A neural network does not offer this ability, and therefore is not a very popular model for this task even though a deep neural network might give superior predictive accuracy.

Utilizing purely data-driven models without insight into a system can lead to poor assumptions and outcomes. We demonstrate this idea using an example of Markov models in credit scoring. Parts of this example are condensed from (Ho, Thomas, Pomroy, & Scherer, 2004).

A Markov model is based on the assumption that all information needed to predict the future is contained in the knowledge about the present and that the past does not matter. Mathematically, knowledge about the past, present, and future can be represented by a state. A discrete Markov chain assumes that there is a state space $S = 1, 2, \ldots, M$, where M is the total number of states. Let $X_t = i$ indicate that the system is in state i at time $t = 1, \ldots, T$. The Markov assumption assumes that

$$P\left(X_{t+1} \mid X_1, X_2, \ldots, X_t\right) = P\left(X_{t+1} \mid X_t\right)$$

A system is said to be Markovian if the probability distribution of the next state is only dependent on the current state. Specific state transitions are represented by

$$P\big(X_{t+1} = j \mid X_t = i\big) = p(i,j)$$

These individual transition probabilities can be accumulated into a transition matrix for easy and efficient modeling. If the transition probabilities do not change over time, the Markov chain is said to be stationary. If the values can change, the nonstationary transition probabilities are represented by $p_t(i,j)$.

A first-order Markov chain follows the Markov assumption as outlined above. Higher-order Markov chains incorporate more information from the past into the future prediction. The next state in a second-order Markov chain is dependent on the current state and the previous state $P(X_{t+1} \mid X_t, X_{t-1})$, and a third order Markov chain is dependent on the current state and the two previous states $P(X_{t+1} \mid X_t, X_{t-1}, X_{t-2})$. These higher-order Markov chains essentially build more history into the model but have the significant drawback of a larger transition matrix. The transition matrix for an kth-order Markov chain contains M^k elements.

In practice, the transition probabilities must be estimated from data. The maximum likelihood estimate for a transition probability is

$$\hat{p}(i,j) = \frac{n(i,j)}{N(i)}.$$

where $N(i)$ is the number of times state i appears in the data set, and $n(i,j)$ is the number of times a transition from state i to state j occurs in the data set. This can be easily extended to higher-order chains and nonstationary chains.

Markov chains are used to model a variety of systems, and often the Markov assumption is made to decrease complexity without verification. A likelihood ratio test can be used to verify that transition between states are, in fact, Markovian. The null hypotheses for this test assuming a first-order stationary Markov chain is

$$H_0(i): p(1,i,j) = p(2,i,j) = \cdots = p(M,i,j), \quad j = 1,\ldots,M$$

where $p(h,i,j) = P(X_{t+1} = j \mid X_t = i, X_{t-1} = h)$. The log of the likelihood ratio test is

$$X^2 = 2 \sum_{j=1}^{M} \sum_{k=1}^{M} n(k,i,j) \log\left(\frac{n(k,i,j) N(i)}{N(k,i) n(i,j)} \right)$$

and this has χ^2 distribution with $(M-1)^2$ degrees of freedom. A Pearson goodness of fit test can also be performed, which compares the observed counts with the expected counts, to assess the Markovian nature of the entire transition matrix. In many applications in the literature, the Markov assumption is made without performing this test. This can lead to poor models that underperform and do not appropriately approximate the system.

In (Ho et al., 2004), the authors present an example for modeling credit scores with a Markov chain. First, a 10-state stationary first-order Markov chain is estimated from the data. Both tests show that the estimated transition matrix is not Markovian. A reduced model, which groups the 10 states into 3 states, also fails these tests. If these tests were not performed, most practitioners would have blindly made the Markov assumption and moved on to modeling the system. It is difficult to estimate the problems that could be caused by using these models in practice, but it is safe to say that making assumptions about models that are false could be disastrous. One only needs to look at the financial crisis of 2008, where housing defaults were assumed to be independent when they were not, to see how dangerous this can be (Saunders & Allen, 2010).

If one takes a step back and thinks about the credit scoring problem, is it reasonable to assume that a person's future credit state is only dependent on their current state and is completely independent of their credit state history? A person with historically good credit could be in a bad credit state due to an unforeseen circumstance and could be trying very hard to change that state. Further, is it reasonable to assume that transition probabilities do not change over time? A person's behavior could certainly change over time.

The example in Ho et al. (2004) goes on to show that a higher-order model, which segments states in a specific fashion, does pass the Markovian test. This aligns with insight into the problem because it does seem logical that a person's credit history and current credit state would affect their future credit state.

This conclusion also sheds light on another problem often faced when using Markov models, how do you decide on an appropriate state space and how do you map collected data into this state space. Tool wear is another problem often modeled with a Markov chain. The wear on a tool can be measured in a continuous space and must be mapped into a discrete space to use a discrete Markov model. The problem is, given M states (we will discuss selecting the number of states shortly), how do we pick the boundaries between wear states? One very simple method is to select boundaries so that each state has an equal range in the continuous space. While this method may seem logical, it could not align with the physics of the problem. Generally, the first stage of tool wear is a break-in period where the change in wear is very high over a short period. Next, the tool enters a period of low wear and can remain in this state for several time periods. At the end of the tool's life, it enters a failure state where the wear rapidly increases over a short period. Figure 12.5 displays a generic wear curve with the break-in, steady-state, and failure regions indicated. Evenly distributed wear boundaries might not capture this type of behavior. For example, a state's boundaries might be placed such that the transition between the break-in stage and the steady-state stage occurs in the middle of

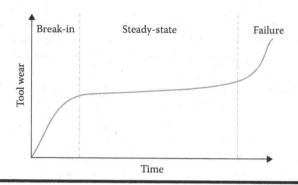

Figure 12.5 Generic wear curve.

the state. This type of insight into the physics of the problem can lead to a better method for selecting state boundaries.

In the previous paragraph, we made the statement "given M states." Many problem formulations presented in the literature on Markov chain applications begin with a similar statement. The number of states is not always readily available. In the credit scoring example, the presented data suggests 10 states; however, the optimal model segments these states into subgroups. In some problems, the data would suggest a very large or even infinite state space (the continuous tool wear data could suggest an infinite state space). Several approaches have been presented in the literature for selecting the optimal number of states given a data set, but the state boundaries must be defined. These methods often evaluate the trade-off between model accuracy and model complexity. In our experience, only insight into the data, the problem at hand, and the goal of the decision system can lead to adequate solutions for selecting the number of states and the state boundaries. Even then, these are only models and representations of reality. There is no correct or optimal answer to any of these questions. There is only a practitioner's best guess guided by insights and intuition.

Conclusion

In this chapter, we have highlighted the need for additional knowledge that needs to be part of the machine-learning paradigm. We have coined this approach as machine learning in context; that is, the framework of the problem must include, in addition to the data and a constructed predictive model, the insights from the problem domain that effect the decision for which the model was constructed. We've illustrated this with multiple examples that highlight the difficulties that can occur without the proper consideration of the machine-learning context.

References

Adams, S., Beling, P., & Cogill, R. (2016). Feature selection for hidden Markov models and hidden semi-Markov models. *IEEE Access*, *4*, 1642–1657.

Angelopoulus, N., & Cussens, J. (2005, August). *Exploiting informative priors for Bayesian classification and regression trees*. 19th International Joint Conference on Artificial Intelligence, Edinburgh, Scotland, pp. 641–646.

Blum, A. L., & Langley, P. (1997). Selection of relevant features and examples in machine learning. *Artificial Intelligence*, *97*(1), 245–271.

Coleman, M. C., & Block, D. E. (2006). Bayesian parameter estimation with informative priors for nonlinear systems. *AIChE Journal*, *52*(2), 651–667.

Dash, M., & Liu, H. (1997). Feature selection for classification. *Intelligent Data Analysis*, *1*(3), 131–156.

Gauvain, J. L., & Lee, C. H. (1994). Maximum a posteriori estimation for multivariate Gaussian mixture observations of Markov chains. *IEEE Transactions on Speech and Audio Processing*, *2*(2), 291–298.

Gibson, J. E., Scherer, W. T., Gibson, W. F., & Smith, M. C. (2016). *How To Do Systems Analysis: Casebook and Primer*. Hoboken, NJ: Wiley.

Guikema, S. D. (2007). Formulating informative, data-based priors for failure probability estimation in reliability analysis. *Reliability Engineering & System Safety*, *92*(4), 490–502.

Guyon, I., & Elissee, A. (2003). An introduction to variable and feature selection. *The Journal of Machine Learning Research*, *3*, 1157–1182.

Ho, J., Thomas, L., Pomroy, T., & Scherer, W. (2004). Segmentation in Markov chain consumer credit behavioural models. In L. Thomas, D. Edelman, & J. Crook, *Readings in credit scoring: Recent developments, advances, and aims* (pp. 295–307). New York, NY: Oxford University Press.

Jang, H., Lee, S., & Kim, S. W. (2010). Bayesian analysis for zero-inflated regression models with the power prior: Applications to road safety countermeasures. *Accident Analysis & Prevention*, *42*(2), 540–547.

Jaynes, E. T. (1985). Highly informative priors. *Bayesian Statistics*, *2*, 329–360.

Lord, D., & Miranda-Moreno, L. F. (2008). Effects of low sample mean values and small sample size on the estimation of the fixed dispersion parameter of Poisson-gamma models for modeling motor vehicle crashes: A Bayesian perspective. *Safety Science*, *46*(5), 751–770.

Mukherjee, S., & Speed, T. P. (2008). Network inference using informative priors. *Proceedings of the National Academy of Sciences*, *105*(38), 14313–14318.

Rabiner, L. (1989). A tutorial on hidden Markov models and selected applications in speech recognition. *Proceedings of the IEEE*, *77*(2), 257–286.

Raina, R., Ng, A. Y., & Koller, D. (2006, June). *Constructing informative priors using transfer learning*. Proceedings of the 23rd International Conference on Machine Learning. ACM, New York, pp. 713–720.

Saunders, A., & Allen, L. (2010). *Credit risk management in and out of the financial crisis: New approaches to value at risk and other paradigms*. (Vol. 528). Hoboken, NJ: John Wiley & Sons.

Thomas, D. C., Witte, J. S., & Greenland, S. (2007). Dissecting effects of complex mixtures: Who's afraid of informative priors? *Epidemiology*, *18*(2), 186–190.

Vanpaemel, W. (2011). Constructing informative model priors using hierarchical methods. *Journal of Mathematical Psychology*, *55*(1), 106–117.

Washington, S., & Oh, J. (2006). Bayesian methodology incorporating expert judgment for ranking countermeasure effectiveness under uncertainty: Example applied to at grade railroad crossings in Korea. *Accident Analysis & Prevention, 38*(2), 234–247.

Winkler, R. L., Smith, J. E., & Fryback, D. G. (2002). The role of informative priors in zero-numerator problems: Being conservative versus being candid. *The American Statistician, 56*(1), 1–4.

Yu, S. (2010). Hidden semi-Markov models. *Artificial Intelligence, 174*(2), 215–243.

Yu, R., & Abdel-Aty, M. (2013). Investigating different approaches to develop informative priors in hierarchical Bayesian safety performance functions. *Accident Analysis & Prevention, 56*, 51–58.

Chapter 13

Flipping the Script: Key Conversations to Understanding the Business, Science, and Art of Fundraising and Its Synergies with Data Science

James Wing-Kin Cheng

Contents

While thinking about this chapter, conversations have started to replay in my mind relating to fundraising/philanthropy and data science, in which I have grouped into three categories. One set of dialogue revolves around a survey of this "Third Sector," discovering its various verticals, constituents, and processes. Another group of conversations, the crux of this chapter, dives more deeply into the application of data science in philanthropy from data management to data analysis to data communication. Finally, more recent dialogues cover the art of dealing with audiences neither versed nor immersed in data science. On a grander scale, talks also swirl in my head relating to the implications of local, regional, national, and global changes on evolving charities, as well as their relationship to their donors, with regard to everything from the ethics of privacy to diversity and inclusion, both within the charities and with donors themselves. And since I can only think of these ideas as conversations, the following fictitious dialogues detail the business, science, and art of philanthropy in conversational form. Of course, the names have been changed to protect the innocent . . . and not so innocent.

The Business of Fundraising: Under the Hood of the "Third Sector" (as Inspired by *The Office* Fan Fiction)

Jim: Good morning, Michael! There's a stack of mail for you, mainly from charities for rabies education and treatment.

Michael: What? What sorcery is this? How do they know about my passion for rabies?!?

Jim: Only if you consider things like marketing and good data science as magic. One possibility could be that one or more of these organizations bought an acquisition list with your name in it as people with potential interest in their causes. Another possibility could be someone donating in your name to one of these organizations in honor of your fun run for rabies. Folks at these organizations with good data management could simply pull up all their honorees and send you educational material, to see if the honorees respond to these targeted communications with a gift.

Michael: Hmmm, I'm intrigued. It sounds like these nonprofits run like businesses!

Jim: Well, you can draw a lot of analogy between the world of commerce into the philanthropy sector. For instance, there are what can be considered "vertical markets" that make up this sector, as nonprofit organizations have missions that meet various goals. For instance, not-for-profit primary and secondary schools, as well as technical/trade schools, colleges, universities, and other educational institutions comprise the philanthropy sector's "education vertical." Hospitals, medical research centers, and other health-related 501(c)(3) organizations fall under the "healthcare vertical." Oh, by the way, all nonprofit, public charities, as well as private foundations, are designated by the IRS as 501(c)(3) organizations.

Michael: Okay . . . . [Looks into the camera in bewilderment.]

Jim: I'm just getting started! Places of worship like churches, synagogues, mosques, and temples, as well as para-religious organizations, make up the "religion/spirituality" vertical. There's also a "membership" vertical. These can be physical places that folks can go like museums, galleries, and performing arts centers, but they can also be cause-related like political organizations, environment-related organizations, and organizations related to social justice or social services. But, of course, these are somewhat arbitrary categories. Think of a big university that has a medical center, an art gallery, a performance center, and an athletic complex/stadium on top of its various research centers and academic departments. Or how about religiously affiliated schools. There is definitely intermingling among the verticals.

Michael: But Jim! I'm all about the people! What about the people?!?

Jim: Ah, so you want to know about the stakeholders?

Michael: Yes . . . . And the potatoes too.

Jim: [Looks into the camera in bemusement and ignores the remark.] I think people within the philanthropy sector can be divided into how or what they give. One group I'd categorize as the "givers of attendance or time." This group would consist of alums or parents of alums for educational institutions, as well as fundraising event participants or volunteers, patrons, grateful patients, or congregants . . . basically anyone who's willing to spend time to help a particular nonprofit organization.

Michael: So, like me, setting up that fun-run for rabies!

Jim: Yes . . . . And the second group of "givers" is a nonprofit organization's donors, the "givers of wealth." Be it in the form of cash, checks, stocks, estate or trust, or even property, a nonprofit's donor base has various "price points" or gift ranges. The highest level of giving for any nonprofit is called the "principal gift," followed by the "major gift," although sometimes it's just simply "principal/major gift." A smaller gift in terms of dollar amount may be considered an "annual gift." There are no set rules in terms of what makes a principal gift, a major gift, or an annual gift. One nonprofit's "large-ish" annual gift may be another nonprofit's major, if not, even principal, gift.

Michael: Wow, so where does the money come from?

Jim: Well, donors can directly give to their charity or charities of choice if you consider that as one "source of revenue." For major and/or principal gifts, folks can also give more indirectly via private foundations or donor-advised funds. Another source comes from those who have left part or all their wealth to a nonprofit in their wills or estate plans. Usually for smaller, annual gifts, donors can choose to be "sustainers" by making more frequent, repeated gifts, be it yearly, bi-annually, quarterly, or even monthly. As an aside, donors can make gifts that are either "unrestricted" or "restricted" to a certain area within a nonprofit like for a specific research topic, or to a certain position like for an endowed professorship. They can also make gifts in honor of someone or others who is/are living or in tribute or in memory of someone or others who have passed on . . . . [Begins to turn his attention away from Michael.]

Michael: But what about us?!?

Jim: There is no "us." [Makes air quotes.]

Michael: No, I mean, the first two groups, the participants and the donors, are like our customers or clients . . . . So, what about us? [Makes circular, encompassing motion.]

Jim: Oh, I see! Well, the staff within a nonprofit, I guess we . . . . No, I mean, they would be considered the "givers of vocation," basically folks who have made careers related to a nonprofit cause or mission. Wow, Michael, I actually think you're catching on!

Michael: I want to start a nonprofit now! Where and how do we start?

Jim: [Sidebars to a conference room and deadpans into the camera.] There is no "we." [Goes back to conversing with Michael.]

Well, if you want to know about the process or the "pipeline." We sort of talked about this at the beginning. First, you must make "first contact," that is, getting information about a nonprofit's potential customers/clients, hopefully potential donors but potential participants who may also end up being donors. To get information on potential donors, or "donor prospects," for first contact, nonprofit staff will usually have to do proactive research, called prospect identification, for principal or major gift prospects, as well as buy acquisition lists for potential annual donors and sustainers. The next step in the pipeline is the process of cultivation. Sometimes, usually for small or annual gift donors, mass mailing, emails, or even social media messaging are good ways to cultivate donors at these gift ranges. However, for major and principal gift levels, fundraisers within a nonprofit may write personal notes, call or even visit their prospects. With the help of researchers to find information about these potential donors, fundraisers would then be able to ask for the right amount and for the right reasons. Finally, when fundraisers can "convert" prospects into donors, they maintain or steward their relationships with their donors by keeping them up-to-date about how the impact of their gifts . . . as well as to gauge if or when they'd like to give again . . . Hey, Michael, where are you going?

Michael: I'm going to start cultivating my principal gift contacts now for our nonprofit! You can text me all your research for us! [Runs out of the office and toward the elevator.]

Jim: [Stands up and screams.] BUT THERE IS NO US!

The Science of Fundraising: Symbiosis between Philanthropy and Data Science (as Inspired by *Star Trek: The Next Generation* Fan Fiction)

Will: So, Deanna, tell me again why we're back toward the beginning of Earth's twenty-first century to study, of all things, fundraising, for this mission?

Deanna: I think this is a pivotal moment in Earth's history where something as "nonscientific" as philanthropy and something as technical as data science can work so well together as to be symbiotic.

Will: Being assigned as a gift officer for this mission, I've learned a lot about the relational/people aspects of fundraising. I just don't see how data science fits here.

Deanna: It's a good thing that I'm your fundraising data scientist on this mission, Commander.

Will: Imzadi.

Deanna: Oh, please, Will. Stay focused! So, you want to know how data science contributes to fundraising? The first step is to understand what exactly data science is.

Will: Okay, let's do this!

Deanna: If we take data to mean any sort of information that can be stored, data science is the empirical way in which information can be managed, analyzed and communicated.

Will: Sounds simple enough.

Deanna: Yes, we can easily define these three aspects or pillars of data science: data management, data analysis, and data communication. However, each of these aspects or pillars can be seen as a specialty within data science that individuals can concentrate on as their career. For instance, if we look back into Earth's twentieth century, data management at that time was all the rage, as people in this field concentrated on ways to get faster and faster retrieval of more and more diverse types of data. During that time, data management was better known as database administration or report building. As data management evolved, the term "business intelligence" came into vogue as these database-dependent reports became more insightful. When all was said and done, however, at that time, database administrators/data managers were most concerned about how to move and manipulate their data from location to location as quickly as possible. You can see that by another term that was also in vogue at the time called "ETL," which stands for "Extract, Transform, Load."

Will: Okay, great. Now how do the other two pillars of data science come into play?

Deanna: Well, as the volume, variety, and velocity of data increased at a faster and faster rate, people began to wonder about the data's accuracy or, to use another "v" term, "veracity." People also began to wonder, not just how true the data had to be but also how to use all the data available to them in their vast and, oftentimes, expensive databases.

Will: And this is where data analysis comes in?

Deanna: I do believe you're not just another handsome face, Commander.

Will: Compliments will get you everywhere, Counselor, but, please, go on.

Deanna: As Earth moved into the twenty-first century, people needed to justify the data's value for storage. People weren't satisfied with business intelligence reports that gave aggregated results like means, medians, and modes anymore. They wanted more sophisticated analyses. They also wanted things to be more visual and more interactive.

Will: Ah, so the phrase "sophisticated analyses" means . . . .

Deanna: . . . Means, not just providing summary statistics, but analyses that are descriptive or predictive or prescriptive in nature. Here, we have morphed mere data analysis into yet another term in vogue: analytics. More specifically, "descriptive analytics" take historical data beyond just summarizing them as measures of central tendencies. More sophisticated descriptive analyses began recognizing underlying factors/components or segments within your data.

If you think of data as information being displayed as a matrix of rows and columns, where each row represents an individual and each column usually represents a variable, that is, information related to the individual. Factor analyses and component analyses can describe how variables are related to each other in such as way where one factor or component can represent multiple variables. You can imagine, then, where data made up of hundreds or even thousands of variables can be reduced to a much smaller, and hopefully more manageable, number of factors or components.

One great example is the conversion of text into quantitative data for "text analytics" to take place. More and more nonprofits are capturing open responses online as constituents describe their amazing (or not so amazing) experiences with the nonprofits as open-responses during their event registration or gift donation process or as constituents write on different social media platforms. Data scientists can then break down these texts into many "tokens." Be they single words or word phrases, these tokens become dichotomous variables of "0s" and "1s" where factor and component analysis can undercover various "themes" to a group of responses. Other factor/component analytic techniques may be able to deduce a response's ultimate "sentiment" as either positive or negative.

Whereas factor/component analysis summatively describe the columns or variables in your data, the rows of your data can be summarized or reduced similarly through techniques collectively known as "cluster analysis." Because rows in a spreadsheet of data usually represent individuals, cluster analysis can group individuals with similar characteristics into different segments or clusters. Again, in this way, data containing hundreds or thousands of individuals can be reduced or summarized into a more manageable number of segments or clusters.

Will: Just to reiterate, that factor/component analysis, as well as cluster analysis, are data reduction techniques aimed at describing large amounts of data. And I think I can understand why these descriptive analytics techniques are important. Factor analysis and component analysis may point toward commonalities like some underlying construct among multiple variables, while cluster analysis can group individuals into segments or clusters and provide information about the variables used to group these individuals. Both are generating insight into your data by uncovering patterns that would otherwise be hidden within the massive amount of information. But, like our time-traveling mission, we're still just dealing with the past.

Deanna: Ah, yes, Commander. And this is where we talk about predictive analytics, tools, and methods that take historical data and try to predict or forecast the future based on the past. One such example is time-series analysis used in this century's economic market indices, where past market performances are used to predict future market performance at different moments into the future, be it a month from now, 1 year, 5 or 10 years.

When you have some sort of outcome you want to predict, there are many techniques in what's called regression modeling for you to use, depending on what your data and your outcome look like. For instance, if you're trying to predict earnings of companies with company-related variables, you can use what's called multiple regression, which basically creates a mathematical model where predicted company earnings are on one side of an equation and a set of company-related variables used to generate these predicted company earnings are on the other side of the equation.

But there are so many variations to regression modeling. Say your outcome variable happens to be a dichotomous one, like whether you're a paying customer or not. Then you would use a regression technique called logistic regression, where the mathematical model is geared toward predicting a binary or dichotomous variable. There is what's called a multinomial logistic regression, when you want to predict for an outcome variable with more than 2 categories. You can even have a multivariate model where you're predicting for more than one outcome variable.

Will: Fascinating!

Deanna: There's a bit of nuance here. While time-series analysis is used to forecast future events, regression models can be used both for prediction and for classification. Oftentimes, regression model results provide the likelihood that an event or a behavior will occur at the individual level. However, regression model results may also provide the likelihood that an individual will belong to a membership group.

And this is just a few of the many techniques for predictive analytics! From improvements on regression models to more "generalized linear models" and even more advanced classification techniques like support vector machines, predictive analytics create algorithms based on past data, so that predicted outcomes can be calculated when new data are entered.

Will: Ah, but what if the outcomes, the data themselves, and/or the conditions change over time?

Deanna: Yes, this is where prescriptive analytics come into play. Okay, so descriptive analytics provide insight to past performance based on historical data, while predictive analytics provide insight on future opportunities. Prescriptive analytics take both descriptive and predictive analytics results into simulations and scenarios. Thus, prescriptive analytics provide insight for business decisions under various conditions, showing the optimal performance to various "what-ifs."

Will: Having been around other fundraisers during this mission, though, I can tell you that, whatever insights descriptive, predictive, and prescriptive analytics provide, most of us will need a less "analytics" way for these insights to be given to us.

Deanna: Absolutely. I think that's why "data communication" is just as important of an aspect to data science as data management and data analysis. This is where storytelling becomes vital. A great data communicator must take all the insights gleaned from analytics into a digestible form for those who are not data-scientists without diluting the message, so that the call for action may be equally impactful across all audiences. There is a tragic lesson learned about one of Earth's earliest space explorations involving the space shuttle Challenger.

Will: I remember that from my twentieth century Earth history briefing prior to our mission. It had something to do with "O-rings" malfunctioning at low temperatures, correct?

Deanna: Yes, exactly. However, the way the data were presented to decision makers in the space shuttle launch was mired in technicality, such that the recommendation by engineers NOT to launch the Challenger was overridden by the project managers who would have sided with delaying the launch if the information was presented more clearly. Of course, this is an oversimplification of a tragic loss of life in the name of space exploration. However, this points toward the need for clear and impactful data communication.

I think that's why the current technological advances of Earth's twenty-first century have allowed for "data visualization" as a field where pictorial representations of data are conveyed through "infographics." But along with presenting clear stories through pictorial representation, data communicators must be great storytellers as well. I think that's why there's a shift from people calling themselves "consultants" in the last twentieth century to calling themselves "analysts" in the early twenty-first century. When all are said and done, evidence-based decision-making must be based on evidence-based storytelling.

Will: So far, I can see how data science can benefit the corporate world. But I don't have to be Betazoid to know that you can also tell me how data science helps the philanthropy sector.

Deanna: Quite right, Commander. For one thing, when a nonprofit organization experiences growth, be it in event participants, in donations or in staff, the organization will also have more data. Here, given that nonprofit resources are usually much more limited than their corporate counter-parts, you're more pressed to ask the question about the utility of your stored data in terms of how much value can be extracted.

And just as companies, have their "shares of market" within their sector, nonprofits have their "shares of pocket" within their verticals. And while many nonprofits share similar admirable missions, they sadly have to compete for donors' limited pocket shares as revenue. Thus, finding potential constituents who will most likely participate, volunteer, and/or give becomes paramount.

And this is where data science becomes most utilitarian. The organizations that can most easily extract, transform, and load stored data . . . . the organizations that can generate the most insightful descriptive, predictive, and prescriptive analytics . . . . and the organizations that can most effectively communicate their insights to their decision makers and even constituents . . . . these would then become the most successful nonprofit organizations. ♦

Will: Let's get into the thick of the nebula. How would we use data science along the "philanthropic pipeline?"

Deanna: If we go to the very beginning of the fundraising process, we have to ask ourselves: "What are our goals here? What are we trying to do?" One of the buzzwords in philanthropy at this stage is "acquisition," that is, the conversions of individuals not associated with a nonprofit versus ones that are. More traditionally, organizations in need of new constituents, be they customers for a for-profit group or event participants, volunteers, and donors for a nonprofit organization, may rely on an outside vendor and aggregate level data, to generate acquisition lists of potential con-stituents. However, this could be rather expensive for a nonprofit.

Instead, a nonprofit may be able to select for initially "cold" individuals who became constituents to the nonprofit, perhaps folks who were acquired through an outside vendor in the past and/or individuals who have little to no relationship with other individuals in the database. Descriptive analysis can be simply looking at group mean differences between these individuals and individuals who are most loyal or engaged with the organization. In a more sophisticated fashion, we can use one or more cluster analytic technique, to see how well these individuals cluster or segment together from the rest of the database. Using predictive analytics, a regression model can be generated with these "cold converts" as the outcome. In both instances, we're neither classifying nor predicting for any individual. Instead, we would actually consider the variables that contribute to the cluster solution or that comprise the regression model, to see the most important factors that distinguish this group of individuals from the rest of the database. With this insight, more targeted communications can be sent to individuals outside the nonprofit's database who are most like those "cold converts."

Will: That's great, Counselor. BUT what my fundraiser colleagues and I would be most interested in is potential donors or donor prospects at the top of the giving tier. Can data science help with finding these folks?

Deanna: Ah, that's the next stage of the philanthropic process is "prospect identification." Traditionally, this process solely relied on nonprofit staff specialized in "proactive research," to gather publicly available wealth information, upon which, total wealth, as well as individual "giving capacity," can be estimated. The problem lies in the fact that proactive researchers cannot provide wealth and giving capacity estimates for everyone in the database. Thus, the goal for data science at this stage is to provide proactive researchers only a fraction of the database of unrated individuals who could potentially have high wealth and giving capacity.

Similar to the data science techniques at the acquisition stage, business rules can be coded into a nonprofit's database, so that unrated individuals that share characteristics similar to high wealth/giving capacity individuals because of some descriptive analytics can be identified automatically and regularly. Predictive models can also be built using individuals with known wealth or giving capacities at and above a certain threshold as the outcome. Such models can then provide probability scores for all individuals in the database, to identify high wealth/giving capacity "suspects." The nuance at this stage is that there might be different types of high wealth/giving capacity individuals; some sort of cluster analysis may be required before any predictive analytics can be applied. And given a clear, succinct way to communicate descriptive and predictive analytics results, proactive researchers would hopefully be pointed into the right directions toward high wealth/giving capacity individuals.

Will: Hmm, and I'm assuming these techniques aren't relegated to find those with high wealth or giving capacity like principal and major gift donors, right?

Deanna: That's right; it really depends on how characteristically different your outcome group is from the rest of the general population in your database. Oftentimes, nonprofits with any sort of degree of data science capabilities will develop predictive analytics for even "leadership" level for annual gift donors, or the highest levels of giving before being considered a major gift. Predicting for planned gifts, or gifts left by donors in their trusts or wills, is also another area where data science techniques are being used beyond just estimating wealth and giving capacity at the principal and major gift levels.

Will: Interesting. I'm just thinking about predictive modeling for a minute here. If your models are based on past and current donors, are you confounding the donors' affinity for your organization with their actual giving?

Deanna: Yes, but I think this is where the end justifies the means. Does it matter how you identify donor prospects for your researchers and ultimately your fundraisers? With that said, you can create separate affinity models, to identify constituents with the potential of having great love for your organization, where these individuals could become great advocates, volunteers and/or event participants for your organization. For nonprofits, however, the best resource is usually financial, in terms of fulfilling their respective missions.

Will: But there's still a relationship component to this, right?

Deanna: Of course, and here, data science can help too! One of the best predictors, if not THE best predictor, of future giving to any nonprofit organization is past giving. Beyond having sustainers, or individuals who have automated making monthly donations to a nonprofit, one of the best ways to increase the chance of donors making repeat gifts would be to maintain and "steward" the relationship between a nonprofit and its donors. Obviously, beyond actual fundraising, another major function for fundraisers and other gift officers or staff members is donor stewardship by keeping regular contact with donors and keeping donors interested in the organization's mission and direction. But then, the question is "Do all individuals have an equal chance of making repeat gifts?" Again, similar to acquisition, prospect identification, wealth estimation, giving propensity, and engagement/affinity, data science techniques can be used, to predict and identify first gift donors who are highly likely to give repeat gifts.

Will: Hmm, I can see how if someone hears us talking about using data science for fundraising, she or he would feel a bit "creeped out." This would seem to be the epitome of being impersonal to the point of leaving potential constituents feeling like, to use an ancient American anachronism, "widgets" in a system.

Deanna: Understandably, if a nonprofit's constituents only experience the organization's facts, figures, and "the bottom line," then they'd feel as if they were in some sort of well-oiled Ferengi profit-making scheme. That's where the staff within a nonprofit must keep both the organization's mission and its people in mind. Some, such as gift officers like yourself and others, must keep good relationships with our donors through visits, phone calls, personal mail, or even simpler bulk, direct mail, or e-blasts. For data science staff such as myself, keeping a nonprofit's constituency in mind translates into treating everyone's data with the utmost care and respect, not seeing "widgets," but rather, seeing the people behind their data.

Will: Well said, Counselor. Thank you for the superb overview of the science of fundraising. How about we grab some drinks? The beverages in this century are much superior to even the synthehol on our ship.

Deanna: Thanks, Commander, but I've got a date with Data.

Will: I hope we're just talking about what you'd find in a fundraising organization's database!

The Art of Fundraising: Embracing Past and Progress (as Inspired by Recent Alternative Facts)

Me: Good morning, sir.

Don: Good morning. Say, you're from CHINE-na, right? You must be pretty good with numbers . . . so why aren't you working in the stock market or a hedge fund or even a tech startup where they need people like you?

Me: Technically, I was born in Hong Kong when it was still a British colony. I was actually a citizen of the British Commonwealth until I was naturalized as an American citizen when I was in high school.

As for being good with numbers, thank you. I actually don't think I'm that good; just ask my friends when I try to figure out how to split our restaurant bills! However, if we're talking about data science, I positively think that anyone, no matter from what country they were born, who has a knack for wrangling and making sense of messy, disparate, and alternative information, to find hidden truths, and then be able to build convincing AND evidence-based stories to address complex inquiries . . . I think people with these skill sets are what make data science great!

As for being a data scientist in the philanthropy sector, I think one major reason for me is that the challenges faced by data scientists in nonprofits are very similar to those faced by data scientists in for-profit corporations. Beyond all the technical similarities in database technology, statistical methodology, and visualization rules of thumb, I think data scientists in philanthropy and in the corporate world must convince "the old guard" or "conventional wisdom."

Traditionally, for example, fundraisers often rely on current donors' leads for prospects in what survey researchers would call "snowball sampling." While this can still be the case today, competition for donors' attention and eventual share of pocket unfortunately has made traditional ways of prospect identification less effective. As another example of tradition or convention, proactive researchers used to read through newspaper, business, and trade magazines in hopes of finding prospects sympathetic to their organizations' causes. In both cases, you're dealing with a small number of individuals to which other nonprofits may want or already have access.

What if you can provide that list of donor prospects on a regular basis for proactive researchers to confirm wealth and giving capacity or else for fundraisers directly to start building those relationships? With data science, you can do just that! When you have a database of thousands or even millions of individuals with information that other organizations may not even have, so many possibilities exist! You can set up business rules within your database to extract prospects who look similar to different characteristics within your current donor population. You can even create models that predict for these individuals within your database.

The point is that data science can direct proactive researchers and fundraisers in a more efficient manner toward donor prospects rather than simply chasing down word-of-mouth leads or reacting to news or business journal articles.

Don: Hmm, you mentioned something about having my information in your database. Yeah, I don't like that. I don't want people to know so much about me and tell it to the world.

Me: Well, sir, I'm sorry you feel that way, but you do raise a great point about the ethics that all fundraising staff, from the fundraisers themselves to fundraising data scientists, share. One of the utmost important ethics for any fundraising staff is confidentiality of information. We often take "anonymous gifts" where donors do not provide their information at all. While that's not so good for fundraising data scientists in terms of not having any donor characteristics connected to these gifts, we have to abide by that. [Unfortunately, that also usually means these truly anonymous donors cannot reap the tax benefits of charitable giving.] In most cases, we do have donor information attached to their gifts, and, while it's their choice to remain publicly anonymous or not, we provide the utmost confidentiality.

Personally, one of the first steps in preparing data is to "anonymize" the records, so that full names and address information are not retrieved. I would only have their unique key ID variable (and obviously, that is neither their social security nor credit card number). Instead, I might retrieve titles and salutations as proxies for profession, marital status, and perhaps even legacy. As for address, I might create variables to indicate whether the address is a residential home, a business address, an apartment, or a suite. By comparing a zip code in the address to census and other government data, I can even attach any zip-level data to individual records.

Unfortunately, once I have a list of IDs representing donor prospects through data science processes, I do have to pick randomly a few IDs to find the individuals represented to verify the accuracy of the data science efforts. Eventually, both proactive researchers and fundraisers must know these donor names to estimate wealth and giving capacity and to build relationships. However, along the entire process, constituent information remains confidential, known only to staff members who require the information to move forward the nonprofit's mission (i.e., business purposes).

One more thing about ethics. On the organization level, "transparency" is another well-guarded and upheld principle. The best nonprofits have, or should have, an audit trail in terms of where gifts originated, how they came into the organizations, and finally how and where the gifts were spent. This is mainly due a well-oiled, internal gift processing team. Associated with transparency is data integrity. Here, donors benefit from easily accessible giving history for tax purposes. Data integrity is also crucial for data science, to prevent the worst-case scenario of "garbage in, garbage out."

Don: Ah, yes, integrity. That's how I want to be known. Let's get rid of all the dishonest people, all the garbage. Let's get rid of all . . . .

Me: Before you finish that thought, I want to point out that, just as the methods of fundraising have evolved over time, so have the constituents of the philanthropy sector. Beyond the fact that our constituents have followed the changes in the racial and ethnic makeup of this country, nonprofits have also begun to reach outside of their locales to neighboring states, to other regions of the country, and even to other countries around the world. Combined, what makes the most sense would be to have diverse and inclusive staff members who are able to interact with a more diverse and inclusive constituency.

Donors have also evolved in such a way that many now see their philanthropy as "investment for the greater good." With that perspective, donors have become much more analytical and empirical, using evidence and metrics of success and fiscal responsibility by nonprofits as drivers in their decision to give. In order, to meet this demand, staff across all aspects of fundraising need to diversify their skill sets, becoming more analytical/empirical themselves. Conversely, fundraising data scientists must also be more diverse and inclusive, be it in background (e.g., country of origin) or in skillset (i.e., language nuance to understand text analytics results) . . . Sir, are you alright?

Don: Yeah . . . . Just trying to fit all of this in 140 characters.

Me: . . . Right, good luck with that, sir . . . [finds the nearest wall to rest a weary forehead]

Chapter 14

Looking toward the Future with Cognitive Computing, AI, and Big Data Analytics

Judith Hurwitz

Contents

We are in a revolution and a renaissance in data analytics. While computer scientists have been working on solutions to complex data analytics problems for decades, three factors have converged to enable this revolution:

Factor 1: The ability to collect massive amounts of data. With the ability to collect more data from applications, the Internet, and connected devices, organizations have made it possible to access more data than was ever imagined. Advances in medical devices, connected machinery, and data from massive textual repositories means that there is more information than ever before. There is a sense of urgency in being able to take advantage of this treasure trove of data to make new discoveries and transform industries.

Factor 2: The availability of inexpensive compute and storage. The advent of the cloud with inexpensive compute and storage services have meant that it is possible to analyze huge amounts of complex data at a reasonable price. In the past, analysts were forced to pick subsets of data sets for analysis because of costs and other computing limitations. New storage techniques such as NoSQL and Hadoop have emerged to allow companies to affordably store and manage vast amounts of data.

Factor 3: New analytics techniques and tools have emerged. These tools can abstract the complexity from the tasks of understanding data. Advances in predictive analytics, artificial intelligence (AI), machine learning, and cognitive computing are promising to change how we work and live. Advanced analytics were once a prevue of a small group of specially trained mathematicians and scientists. Now, new analytics tools are abstracting complex algorithms into a framework that is easier for business analyst to use.

Understanding the Wide Range of Options

The reality is that with all the data analytics technological advancements, there are more and more ways that computer scientists can use techniques and tools to advance the way computers can use data for learning and transforming processes and tasks. Many of the resulting solutions will transform industries, the way we work, and the type of solutions to problems that could never be solved. In this chapter, we will provide an overview of what we mean by Big Data and then put this in perspective with cognitive computing, AI, and machine learning.

What Is Cognitive Computing?

Cognitive computing has some distinct differences from traditional computing solutions and traditional analytics approaches. Let's start with a definition: Cognitive computing and the techniques of AI is an approach that enables humans to collaborate with machines. It enables data to be analyzed in context based on a variety of data including text, images, voice, sensors, and video. Cognitive computing is a form of AI in that it uses the ability to have computers perform tasks that would have traditionally be performed by humans. A cognitive system is designed based on three capabilities:

- A cognitive system learns based on the data ingested. Through the learning process, the system makes inferences about the area being analyzed.
- A cognitive system requires a model or a representation of a domain. The model must understand the context of the data being used.
- A cognitive system must be able to generate hypotheses. To make sense of the data, a system is required to come up with an assumption or explanation of an expected outcome.

The elements of a cognitive system have a common requirement: That enough data be ingested, analyzed, and tested to determine whether assumptions can be supported. Therefore, one of the benefits of a cognitive system is that a bias must be supported by data. If a system is fed enough of the right data, it is possible to determine whether a hypothesis holds up to scrutiny.

The value of a cognitive approach to computing is that the solution design is determined by the patterns within the data, rather than by predetermined logic. But in a cognitive system, it is not just data in isolation. A cognitive computing system requires the knowledge of content experts to curate the data and provide insights into the right data sources. Therefore, a cognitive system only works when there is collaboration between human experts and data that supports the way they put their data to work.

What does it mean to create a cognitive application? It's not that much different from creating any application in many ways. You have to start with understanding the objective. For example, you might want an application that will help consumers plan a trip or you want an application that helps anticipate and avoid security problems in software. However, unlike a traditional application development process that is designed to answer specific predetermined questions, a cognitive application is intended to go deeper and explore context. In a cognitive application, the intent is to look at the relationships between data elements by creating a specific domain related to the area of focus.

By defining a narrow domain, you can then begin to determine which data sources are the most important. This data is then put into a corpus—a machine—readable representation of the complete record of a topic. But that is only the beginning of the process. What makes a cognitive system powerful is that the data is not static. The system learns from the interactions with both experts and users. Through an initial testing process, experts determine whether this corpus provides the right context to come up with answers and solutions. If the context is wrong, it is likely that there needs to be new and different data provided. The power of the cognitive system comes from its ability to iterate on the data. Over time, the system learns and changes based on the interactions between machine and human.

The advantage of a cognitive approach to solution development is that it begins with data rather than assumptions about business logic. Through machine-learning approaches and the collaboration between humans and machines, the cognitive system can evolve as the business changes. A fast-changing world requires that organizations understand the hidden patterns and anomalies in data so that organizations are prepared for shifts in customer needs.

Cognitive computing is not a single technology. Rather, it is a combination of a variety of algorithms and techniques intended to help practitioners make sense of the enormous amount of information they have—both structured and unstructured to make better decisions. It has long been the dream of researchers and computer scientists to find techniques to transform data into knowledge. The path to achieve success has been fraught with failures and missteps. Many horror movies were made depicting the ability of humans to create artificially intelligent constructs that would take over the tasks of humans. However, despite many false starts we are finally entering an era where cognitive computing and advanced analytics is becoming a reality.

The Complexity of Learning

While it is easy to get right into the concepts behind cognitive computing and machine learning, it is important to understand why translating human learning into a systematic approach is complicated. Think about how humans learn. Take the example of the five-year-old girl who has been given her first two-wheel bicycle. While she has been very comfortable riding a tricycle, learning to ride a two-wheel bicycle is another matter. First, the child has to ingest a lot of information relating to riding a bike. What are the dynamics of balancing with the two wheels? How fast does the child have to peddle to move forward without falling over? Eventually the child understands that she has to balance on the frame of the bike in the right way so that it doesn't fall over. She has to navigate different surfaces. There are different techniques needed to navigate a bumpy dirt path versus a well-paved road. After a while she can intuit how to navigate without assistance. Riding a bike becomes second nature. How does the child navigate on pavement versus grass or an unpaved dirt path? How does the weather affect her ability to ride? Is it windy or rainy? The child needs some assistance at first to understand what it feels like to stay upright on a bike. However, even with some coaching, the child will have to learn by trial and error. Learning is actually complicated, involving training, experimentation, and constant learning.

Training from Data

Humans have a built-in capability to learn and change. Training a system based on Big Data is much more complicated. Therefore, the question is how do you train a system to gain the same intuition about executing a task? It is clearly not easy. But we are entering an era where we can use advanced analytics approaches to allow us to gain an understanding in context.

Collaboration between Humans and Machines

A cognitive system is a technology approach that enables humans to collaborate with machines to gain insights and take actions in context. One of the principles of cognitive computing is the availability of enough data that can be analyzed at the right level of performance. Only a few years ago, it would be prohibitively expensive to store and apply advanced computational analysis to the data. With the advent of much more powerful processors and cloud computing combined with the dramatic drop in costs for both compute and storage, advanced analytics has made huge advances.

Now that we have the platforms to support this new generation of analytics, we need to understand the steps required to turn data into knowledge. How do you turn information into knowledge in the real world? The key to the practical application of cognitive computing is to focus on specific domains. To gain insights from data means that you have to understand the context.

Trying to understand the patterns and anomalies from data requires that you focus. Therefore, you need to focus on a single domain—such as retail, energy production, insurance, or healthcare. By focusing on a domain, it will be possible to gain an understanding of the foundational data for that field based on best practices and patterns.

The Relationship between Big Data and Cognitive Systems

Creating a cognitive system is predicated on domain knowledge, best practices combined with massive amounts of structured and unstructured data. We call this Big Data for lack of a more precise term. When we talk about the requirements of data as it applies to cognitive computing as having the following four foundational characteristics.

Volume

Volume is the most obvious characteristic of Big Data. However, volume can vary depending on the nature of that data. For example, there may be a massive volume of data from sensors in an industrial machine but the data itself is quite simple. In contrast, the data that comes from a medical image is massive and it is complicated. Unstructured data that comes from thousands of pages of text such as clinical trials or medical journals is quite complex because it has to be translated into a form that a machine can understand.

Variety

To make sense of data in context with the domain being addressed and the problem being solved requires a lot of different types of data. There is highly structured data that you would find in a database. There is semi-structured data that would be found in data generated by sensors, industrial equipment or images. There is also unstructured data that is generated by documents. While we call this unstructured data, it does have internal structure that can be understood—but accessing and interpreting that data is complicated.

Velocity

Velocity refers to the speed of moving data form one location to where it needs to be ingested and stored. The need for speed will vary depending on the nature of the use case. For example, when there is a massive amount of unstructured data from social media sources may require high speed to both move and analyze the contents of the data. In other situations, speed is less important such as when you are ingesting incremental updates to data to continue the learning process.

Veracity

Veracity can often be the most important issue related to Big Data, because if the data isn't reliable it will provide erroneous results. This is not black and white. For example, you may want to ingest a massive amount of unstructured data to begin to test a hypothesis. Some of the data will be very important but other elements of data may be noise. The irony of Big Data is that the results and answers will come from a small amount of data where patterns or anomalies are found.

The Architecture of Big Data

Big Data cannot live in isolation. There is a framework of services including the scalable and adaptable physical infrastructure that is often highly distributed and typically cloud based. As with any mission critical system, the entire system must be managed in a secure manner. There have to be a supporting software infrastructure that allows data to be ingested (from internal and external sources) and integrated, managed, analyzed, and governed. In this chapter, we will not discuss the details of how these elements are architecturally managed. Rather, we will give you a sense of the types of data needed and the types of operational databases needed to support cognitive computing. In this next section, we will address the roles of both structured and unstructured data.

The Role of Structured Data

While most of the discussion around cognitive computing often centers on unstructured textual data, it is equally important to understand the role that structured or semi-structured data plays. Traditionally, structured data has a defined length and format and its semantics are defined in metadata, schemas, and glossaries. Cognitive solutions often need structured data to create the context between the unstructured elements and how they impact the analysis. For example, you may be collecting a massive amount of unstructured data about a specific disease. However, you also need companion structured data that gives you demographic data into the type of patients who contract this disease and outcomes.

The Value of Semi-Structured Data

Semi-structured data is critical for a cognitive solution. Semi-structured data is typically machine-generated data from devices such as sensors, smart meters, and medical devices. Unlike structured data, semi-structured data is not transactional. Rather, this semi-structured data does not have explicit formats and so its interpretation has to be gleaned from applying machine-learning algorithms that translate this data into understandable content. For example, sensor data from a manufacturing machine can provide an early alert system if the machine is overheating.

The Role of Unstructured Data

As the name implies, unstructured data does not have a formal transactional structure. Rather, unstructured data does not follow specific formats like a database would. In addition, the unstructured data does not have explicit metadata, schemas or glossaries. Making sense of unstructured data therefore requires that the context and semantics must be discovered and extracted through different processes including Natural Language Processing, machine-learning algorithms, or text analytics.

Unstructured data is the heart cognitive computing for several reasons. First, as much as 80% of the world's data is unstructured. Second, huge amounts of knowledge and information about a field is encased in documents, customer support systems, images, website content—just to name a few. While we call these data sources unstructured, the reality is that every data element has an innate structure that can be discovered. The following are some examples of the database tools that are foundational to understanding unstructured data:

- There are a variety of tools such as Key-Value Pairs databases that provide pointers to the meaning of data.
- Document databases manage repositories of unstructured data so meaning can be understood.
- Columnar databases manage data efficiently by storing data in columns rather than rows. The data is stored in memory so that information retrieval is much faster for text.
- Graph databases. This database format uses a graph structure with nodes and edges to manage and represent data. It is widely used because it is designed to determine relationships between data without the joining function used in relational databases.

In this next section, we will discuss how we take this variety of data and leverage it to support the emerging cognitive computing systems. With the combination of Big Data and domain-specific cognitive computing and machine-learning algorithms, we can begin to build systems that help us reach conclusions and transform industries.

The Elements of a Cognitive System

A cognitive system is designed to solve practical problems by continuously adapting based on data and experience. Traditionally, businesses have used computer programming to solve problems by beginning with translating processes into a set of steps that can be executed. Once the program is written, data is fed into this model providing answers. This approach has been used for decades and is especially useful when business approaches were relatively stable and where there was a limit to how much data an affordable system could store and compute. However, over time, the approach of building complex system based on aging assumptions has become problematic. Today, organizations are finding that business models that worked well for decades are now being challenged. Emerging companies are creating new innovative business models that are challenging incumbents in almost every industry. This means that application logic must be changed in real time for organizations to be able to compete.

The emerging model for creating business value is predicated on the ability to create new logic based on emerging business models. The only practical way to plan for the unknown is to lead with the data. Therefore, rather than writing logic, emerging applications will be designed based on what the data tells us. This makes sense when you leverage unstructured data that provides insights into best practices and patterns. This model-based approach enables organizations to be able to change by applying advanced analytics to business practice. This approach of leading with data combined with best practices is the wave of the future and the foundation for a cognitive approach to computing. The three most important stages include the ability to learn, to model, and to generate the hypothesis.

Learning

Learning is the foundation of cognitive computing just as it is for human knowledge acquisition. But to learn, it is critical to be able to understand the context between data elements. A very simple example explains what this means. Imagine that you are walking down the street and you see a tree where all the leaves have turned brown and have begun falling. If this takes place during the month of November and you live in a climate where there are four distinct seasons, you will assume that this is normal and the tree is healthy. However, if the same situation takes place during the middle of the summer, you will know that the tree is probably diseased. The same set of conditions will take on entirely different meanings based on context of the data. In this case, the domain is the environment of the local area that would include information about the seasons, weather conditions, horticulture, and the like.

In a cognitive system, there is a requirement to make inferences about the data. However, to get to the point where you can make an inference about what the data means requires that the data be focused on a specific domain such as one topic or issue. For example, the focus might be on traffic management or on a specific disease. Therefore, it is the responsibility of the organization doing the analysis to collect as much information as possible that is specific to the topic area. If the topic area is too broad, it will be difficult to collect enough information to support the type of advanced analytics or learning to solve a problem. The domain needs to have enough complexity that it is worth the investment. If there is sparse data, it is likely that the human will be able to review the data and come up with a quick answer. On the other hand, there are situations in medical diagnosis where there is so much information that a human cannot possibly understand and absorb all the data in context to make informed decisions. For example, a doctor may see a patient that has a set of symptoms that don't match a known cause. There might be skin irritation combined with a fever and sudden weight gain. While each symptom might indicate a specific condition, the doctor might never have seen this combination of symptoms. However, there may be information published in a medical journal three years ago that provides a diagnosis that matches these symptoms. This situation requires a probabilistic type of analysis. In a probabilistic system, there might be a variety of answers to a problem depending on the circumstances and context. Most importantly, there needs to be a significant degree of confidence based on the information that has been collected. In contrast, a deterministic system will return a single answer based on the evidence collected.

Modeling

The model is the centerpiece of a cognitive system. The model is not a single element. The model is a representation of the domain that is being analyzed to discover insights, and predict outcomes. Modeling begins by defining a corpus. The corpus is the body of knowledge to continuously update the model based on both ingesting new data and feedback from knowledge experts.

In addition, the model requires a set of assumptions and algorithms that generate and score hypotheses to answer questions, solve problems, or discover new insights.

The Hypothesis

Before you can come up with answers and patterns from data, you need to focus on your hypothesis. A hypothesis is a testable assertion from evidence that can explain an observed phenomenon or relationship between elements within a domain. The hypothesis must have supporting evidence or knowledge that explains the causal relationship. In essence, a hypothesis has to be able to predict an experimental outcome. In cognitive computing, it is necessary to find evidence to either support or refute the hypothesis. Are the leaves falling from the tree because it is diseased? The collected data is analyzed and tested. The result of that analysis is given a score based on the level of confidence. When dealing with unstructured data it can be difficult to prove a hypothesis. There may be several hypotheses that need to be evaluated and scored in parallel. To get a level of confidence in an outcome, this will require a continuous machine-learning process. For example, you will begin by creating your hypothesis, identifying the data sources and ingesting them. The resulting knowledge base of ingested data is called a corpus. Once data is ingested the evidence and the hypothesis is scored, the results are presented to subject matter experts who provide feedback. Is this an acceptable answer? Does it make sense? If the initial result is positive, more data is ingested and training begins. Training of data is the process of insuring that the hypothesis is provable and generalizable. The result of this learning process is the creation of a model that can be applied to solve similar problems.

Machine Learning

To achieve the ability to transform applications without reprogramming requires continuous learning. At the heart of this process is to apply the ability to acquire, manage and learn from data by applying machine-learning algorithms. Machine-learning algorithms are programs that are designed to look for patterns in data and make recommendations for actions based on findings. Patterns are the most critical element because they indicate a similar structure or value. For example, a human face has patterns indicating eyes, noses, and so on. With enough data, it is possible to distinguish first a human face from an animal and with even more data it is possible to determine a specific individual's face. Cognitive computing systems use machine-learning algorithms based on inferential statistics that is the basis for being able to predict outcomes.

There isn't one type of algorithm that is used in a cognitive system. Rather, depending on the nature of the problem being addressed. In this next section, we will discuss the primary learning approaches that can be applied to cognitive computing including supervised learning, reinforcement learning, and unsupervised learning.

Supervised Learning

Supervised learning is a technique that teaches a system to detect or match patterns in data based on examples encountered during training with sample data. Supervised learning is a process used where you already have an idea of what you are looking for based on an existing pattern. This can also be thought of as learning by example. The power of this type of machine-learning process is that the system uses new data based on the example to improve its performance on pattern matching tasks. Within the area of supervised learning, the job of the algorithm is to create a mapping between the input and output.

Typically, with supervised learning, the analyst starts out with very noisy data that includes a lot of extraneous details that aren't needed for the analysis. Over time with training, the learning process is optimized based on weeding out biases and assumptions. Often supervised learning is used to solve classification (to satisfy constraints) or regression (used to fill in expected values) problems. For example, supervised learning could be used in a travel application to determine the right vacation for a specific client where the goal is to learn enough about what the traveler wants based on potential vacation locations. The algorithm determines the patterns of travelers' preferences. A regression algorithm is designed to determine the value of a continuous variable such as price. If a customer is using an application designed to help them purchase a specific car, a regression algorithm will help match the details of the car with the pricing variables. The desired outcome is the car at the desired price.

Reinforcement Learning

Supervised learning is widely used in machine learning and cognitive computing. Reinforcement learning is a special case of supervised learning where the system gets feedback on its performance to help determine the best outcome. While supervised learning systems are trained with an explicit set of training data, reinforcement learning involves the system taking actions based on trial and error. Therefore, sequences of successful decisions result in reinforcement to help the developer generate an approach that can be generalized to solve a problem in a domain. For example, reinforcement learning is widely used in robotics or in self-driving cars. So, when the car takes an action based on data that results in the car driving into a wall, the system (thankfully) learns to take other actions with better results.

Unsupervised Learning

Unsupervised learning uses inferential statistical modeling algorithms to discover patterns in data. Unsupervised learning is used when you don't understand the patterns in the data but are trying to see what might be hidden within the data. The data is used to discover which relationships among data elements or structures are important through such attributes as the frequency of occurrence, the context of where that data appears, and the proximity of the data. For example, in drug discovery it is important to understand whether there is a relationship between a disease-causing molecule and a data structure that seems to have an impact on the disease molecule. Is the disease molecule changed when a certain treatment is applied? Does the data indicate that the pattern can be repeated? In a discipline like drug discovery, you can't predetermine where a pattern exists until you experiment with different hypotheses. Other areas where unsupervised learning is important include vision or image analysis.

Cognitive Computing in Practice

Although we are early in the evolution of cognitive computing, we are already beginning to see how this approach to computing is changing businesses. Traditionally, businesses have relied on the experience of experts and their ability to keep up with changes in technology and business to keep ahead of the competition. How do you translate huge amounts of human knowledge into a system that allows organizations to analyze data to support decision-making? How is it possible to allow the data to help you anticipate changes so that it is possible to take the right action at the right time to change outcomes?

The inexperienced resident in a hospital visits a patient with symptoms of a condition that he has never seen before, nor has he read it in any of his books.

He knows that something is wrong but what? He goes to various databases and information sources and still can't find the solution. Just in time, a physician with more than 30 years of experience enters the patient's room and is able to determine what disease the patient has within five minutes. The difference is the internal knowledge base of the experienced doctor. If you could capture the data that is inside this physician's head and add in other data sources and the context of his experience, it would be possible for the new resident to have solved the problem. This is the promise of a cognitive system.

Getting away from the standard way we approach problem solving to a cognitive approach supported by machine learning takes time. First, as we discussed earlier, it requires that a system be constructed, data be added, and a system be trained based on the domain and knowledge.

The steps to design a cognitive system require at least seven stages:

1. **Defining the objective.** This means that the organization has to determine what type of problem they are trying to solve. Perhaps the organization wants to build a travel application that ensures that a traveler can find the optimal vacation, make the reservations, and build long term loyalty to the travel company.

2. **Defining the domain.** There are thousands of potential domains ranging from medicine, to financial services, to insurance, to retail and travel. A successful domain will have the availability of a lot of data from a variety of sources that can be combined in different ways to solve problems that are almost impossible to achieve manually.

3. **Understanding the intended users.** Who will benefit from the cognitive system? If there are only a few users, then the system will have limited applicability. However, there are domains that have a large pool of users who can benefit. For example, in retail there are huge numbers of customers who are unable to find the products they want when they want them. Retailers are threatened by new business models that are forcing change. A cognitive system that targets the right buyers can change the fortunes of an industry.

4. **Defining the questions.** What are the questions you would expect users to ask? How can you set up a system that helps provide the insights your users need to take the right action? In a medical situation, you would want the inexperienced resident to be able to ask questions about the symptoms of a disease and explore potential answers. Some answers could be hidden in a journal article written five years ago that the resident would never have seen. There may have been a patient last week with the same symptoms who was successfully treated. Not only do you have to ask the right questions but you should have enough understanding of the context of that question to anticipate the next action. Asking questions and getting answers from a variety of sources can be the difference between success and failure.

5. **Acquire the relevant data sources.** You can design the best system but if the data isn't good or if there isn't enough of the data, you will fail. Therefore, one

of the most important characteristics of a cognitive system is to have enough trained data that the system continues to learn and change.

6. **Create and refine the corpus of data.** Once you have the right data you have to continue to add to that data over time. Just as humans learn with time, data, and experience, so does a cognitive machine-learning system. There must be a collaboration between the data and interpretation by subject matter experts.

7. **Training and testing.** Creating an effective cognitive system requires modeling, development, analysis, training, and testing. You need to be able to measure responses to questions to determine how accurate your answers are. After creating a system that provides accurate answers you will begin to establish ground truth—a set of data that is the gold standard for accuracy. This process is ongoing and iterative if the knowledge base changes and grows. Any domain worthy of an investment in a cognitive system will continually create new data.

The Future of Cognitive Computing

We have only begun to scratch the surface of what will be possible when we apply a cognitive computing approach to the data we already own. Today, there is a lot of trial and error in converting what we understand as humans into systems that learn and support our actions. In the future, cognitive computing will be a set of enablers that allow systems to understand when an error occurs and then repair itself. Cognitive capabilities will be part of any technology that we deploy. We will begin creating packaged solutions including pretrained data sets that can quickly allow us to ask the right questions and get the right answers. At the same time, we will use the sophisticated machine-learning algorithms and tools to create new solutions to problems that we never thought could be understood.

Index

A

Advanced analytics, definition of, 199
Advertising message, central
    elaboration of, 144
AIER, *see* American Institute for Economic
    Research
Alogical intuition, 10–11
Ambient marketing, 152
American Institute for Economic Research
    (AIER), 55, 58
Analytic base table (ABT), 207
Analytics
    block diagram of, 200
    business act on, 210–211
    community, 198
    definition of, 199, 200
    importance of, 200–201
    methodology, 208–209
    problem framing, 207
    process frameworks, 201–202
Anchors of trust, 42–43
    effectiveness, 44
    integrity, 44–45
    quality, 43–44
    resilience, 45
Artificial intelligence (AI), cognitive computing
    and techniques of, 249
Authentic trust, 73, 74

B

"Balloon effect" model, 144
Basic trust, 73, 74
Bayesian statistical analysis, 215, 217
Bayes theorem, 215
Benevolence, in social reports, 126, 128, 129,
    132–133

Big Data

Big Data
    architecture of, 252–253
    vs. cognitive system, 251–252
    structured and semi-structured data,
        253
    unstructured data, 253–254
Black box model, 225
Blind trust, 73, 74
Business problem framing, 205–206
Buzz marketing, 152

C

Calculative trust, 73, 74
Calculus-based trust, 74
Career skills, of contemporary workplace,
    54
Certified Analytics Professional (CAP), 204
Characteristic-based trust, 74
Cognition-based trust, 179
Cognitive computing, 249–250
    Big Data *vs.*, 251–252
    design steps, 259
    elements of, 254–255
    hypothesis, 256
    learning, 255–256
    learning complexity, 250–251
    modeling, 256
    reinforcement and unsupervised
        learning, 258
    supervised learning, 257
    training from data, 251
Cold rationality
    decision-making process, 24–28
    to human imperfection, 23–24
Competence trust, 74
Conditional trust, 74

261